Conflict Minerals in the Democratic Republic of Congo

PETER LANG
PROMPT

PETER LANG
New York • Bern • Berlin
Brussels • Vienna • Oxford • Warsaw

Dylan Scudder

Conflict Minerals in the Democratic Republic of Congo

Part of the Pentalemma Series on Managing Global Dilemmas

PETER LANG
New York • Bern • Berlin
Brussels • Vienna • Oxford • Warsaw

Library of Congress Cataloging-in-Publication Data

Names: Scudder, Dylan, author.
Title: Conflict minerals in the Democratic Republic of Congo: part of the Pentalemma series on managing global dilemmas / Dylan Scudder.
Description: New York: Peter Lang, 2020.
Includes bibliographical references.
Identifiers: LCCN 2019041661 | ISBN 978-1-4331-7561-9 (hardback: alk. paper)
ISBN 978-1-4331-7562-6 (ebook pdf)
ISBN 978-1-4331-7563-3 (epub) | ISBN 978-1-4331-7564-0 (mobi)
Subjects: LCSH: Mines and mineral resources—Political aspects—Congo (Democratic Republic) | Mineral industries—Corrupt practices—Congo (Democratic Republic) | Mineral industries—Economic aspects—Congo (Democratic Republic) | Business logistics. | Social responsibility of business.
Classification: LCC HD9506.C75 S29 | DDC 338.2096751—dc23
LC record available at https://lccn.loc.gov/2019041661
DOI 10.3726/b16467

Bibliographic information published by **Die Deutsche Nationalbibliothek**.
Die Deutsche Nationalbibliothek lists this publication in the "Deutsche Nationalbibliografie"; detailed bibliographic data are available on the Internet at http://dnb.d-nb.de/.

This publication has been peer reviewed and meets the highest quality standards for content and production.

29 Broadway, 18th floor, New York, NY 10006
www.peterlang.com

I cannot teach anybody anything. I can only make them think.

—Socrates

Contents

Figures

Preface

Conflict Minerals in the Democratic Republic of Congo provides readers with a hands-on approach to dealing with the complexities unique to international management. The book draws strategically on this complexity to help readers cultivate greater critical thinking. It goes beyond the usual static business decisions related to setting the right price or memorizing truisms of management, choosing instead compelling real-world cases that require detailed deliberation and creative problem-solving. While the book has been fictionalized, it draws on first-hand fieldwork and rigorous research on actual business cases.

Each of the three chapters starts with a brief background of the context and the company, and then confronts the reader with a formidable and escalating challenge to work through. The scenarios include a high degree of complexity that companies often face in fragile environments, whether in developing or industrialized countries.

With a focus on cross-sector collaboration and managing the tensions between financial, social and environmental priorities, the book helps readers learn how to develop creative solutions to seemingly insurmountable dilemmas. It contributes in this way to management skills as well as to the knowledge needed to resolve issues in economic development, peace and conflict studies, political science, international relations and related fields.

Although the topics deal with high levels of complexity, they are written in an engaging and accessible way for both current and future managers interested in stakeholder management, business in developing countries, Corporate Social Responsibility (CSR), sustainable business, international entrepreneurship, strategic leadership and global business issues.

Conflict Minerals in the Democratic Republic of Congo is written as a trilogy to let readers experience how a business challenge evolves over three distinct phases in one location. This design enables readers to "stack" their learning from one part of the book to another. As business issues take on an increasingly global character, the material is organized to help the next generation of business leaders prepare for a wide range of scenarios in uniquely challenging environments.

Disclaimer: This book is intended solely as discussion material. None of the content is intended to serve as examples of effectiveness or ineffectiveness of any kind. Though inspired by real-world events, many of the names of the people, groups and other entities appearing in this work have been changed to preserve anonymity. Any resemblance to real persons, living or dead, is purely coincidental.

Acknowledgments

It was Johan Galtung, "Johan" as he preferred, who introduced me to conflict theory as a way of applying creative thinking to seemingly impossible problems, and as a way of making work and life more meaningful. Johan recommended me to an MA program in peace and conflict studies in 1996, which I enrolled in the following year. For the next four years, I traveled extensively with him to mediation interventions, lectures and workshops in Okinawa and across Europe. Johan offered me a position in his NGO, Transcend, and advised me on my master's thesis that explored the Transcend method of conflict transformation. During this time, he also arranged for me to do an internship at the United Nations (UN) in Geneva, which opened for me a rewarding career path at the UN and later in business and academia. Those first years with Johan were some of the most intense, educational and adventurous ones of my life. Each day was filled with hard work, friendship and more humor than two people should ever be allowed to enjoy.

Ian Green was Chief of the United Nations Training Section in Geneva when he hired me as a training coordinator. Ian generously gave me the space to experiment and discover how to create productive learning environments. Though this proved quite costly for him at times, he always went out of his way to make sure I had everything I needed to grow professionally. I can do the work I do today only because of the opportunities I was given in that first stage of my career.

I want to thank Jane Nelson, both for the encouragement she provided me through her book, The Business of Peace, and especially for the personal guidance she gave me. Jane invited me to her office at Harvard University, where she shared with me inspiring insights into the field of business in areas at risk. Much of what I learned from her writings has illuminated for me how ethical decision-making impacts on firm value.

Kiyoshi Amemiya, the inventor of the now-famous demining equipment that transforms minefields into orchards, gave me a once-in-a-lifetime chance to work with him and his team of engineers on a mine-clearing project in Colombia. He provided me with an insider's perspective—even letting me test-drive the mine-clearing equipment to better understand some of the operator's challenges first-hand. The onsite experience helped me appreciate the complexity of working for peace in a way that is also financially sustainable.

In the very volatile days following the Great East Japan Earthquake, it was Ray Bremner OBE, then CEO of Unilever Japan, who offered me a seat at the table where he and other members of the crisis response team would meet to review and plan. This life-changing experience gave me a chance to see some of the leadership behind Unilever's promise of Doing Well by Doing Good in the extreme conditions of disaster relief. This book is very much an attempt to let the reader sit at that same table and join in the decision-making.

Finally, I owe a particular debt of gratitude to my editor and friend, David Hulme. David was helpfully relentless in getting me to think and write more clearly. Visiting David and Satoko for hikes and in-depth discussions at their home in the mountains on the outskirts of Tokyo provided me with the extra push I needed at critical moments along the way.

DYLAN SCUDDER
Tokyo, August 2019

Introduction

This book aims to help business leaders of today and tomorrow learn how to stay profitable in an increasingly volatile world. Drawing on the latest risk management research, *Conflict Minerals in the Democratic Republic of Congo* provides an immersive reading experience based on the real-world case of a global company responding to a highly uncertain environment in the midst of armed conflict. The central challenge concerns how to reconcile the competing demands of diverse stakeholders: those most interested in immediate and long-term profitability, social concerns and environmental priorities. The book is grounded in the insight that such issues are always interdependent. As such, companies that ignore their stakeholders put their own future viability at risk.

This is easy to say, but much more difficult to realize. Following the PDCA model made popular by W. Edward Deming and refined in Japan, the book is organized around the four phases of *Planning, Doing, Checking* and finally, *Acting*. Within this framework, readers develop competencies by working through business challenges that they prepare for, engage in, reflect on and then draw lessons from that can be applied to the next round of challenges they encounter. Readers work as managers at Pentalemma Inc., a major risk advisory that has several large multinational corporations in its client base. *Conflict Minerals in the Democratic Republic of Congo* is part of a series of books on leadership in complex environments, each covering a different location and linked to a specific industry facing its own unique set of challenges. Each book in the series is written as a trilogy with three distinct phases.

The main question for readers is how to select an optimum business strategy from among several that emerge over the course of the session. They receive a series of reports and press releases, which generate a realistic working environment and expose them to modern business concepts. Parallel to this, they are informed by various advisors, such as business mentors or problem-solving specialists. The experience of reading the book is similar to the experience of managing other urgent problems at work, namely, through a flurry of emails and discussions. Immediately after preparing for the upcoming scenario, it's down to business. Rather than the traditional outdoor adventure, classroom lecture or hi-tech animation, readers are embedded in a corporate team and focused on a complex issue that must be solved without delay.

Learning is optimized by immersing readers in high-impact business challenges through intense but entirely plausible scenarios. These might be set anywhere in the world—from the Democratic Republic of the Congo to Japan. What they have in common is the struggle of making critical decisions in the face of uncertainty. Readers must find practical solutions within quickly changing, high-risk situations. By the end of each chapter, readers will have chosen the plan they think their client company should adopt. The outcome, in the form of client feedback, company results and broader social reaction, will be shared with them through review material that describes plausible outcomes to the decisions they made.

In short, the book emphasizes risk awareness and mitigation by demonstrating for current and future managers how planning for the potential consequences of their everyday decisions helps protect them and their stakeholders from unnecessary risk.

Chapter One

Defining the Problem

You are a manager at a company called Pentalemma Inc., which specializes in stakeholder management. The company took its name from the Greek *penta* (five) and *lemma* (assumption), the forgotten wisdom of identifying five solutions embedded within a problem and then choosing the best one.[1] Pentalemma's headquarters are near Geneva, Switzerland, with several hundred employees around the world. The most frequent problem that Pentalemma's clients have is that their stakeholders disagree with each other. Pentalemma's competitors are often limited in their ability to help clients find creative solutions because they tend not to think beyond dilemmas, which offer only two options.

Pentalemma managers offer a better service because they can see five solutions to stakeholder problems. They find the client's most difficult problem, usually a disagreement between their stakeholders, and then create five options for them to choose from. Clients try to choose the best "win-win" option for themselves and their stakeholders.

As a Pentalemma manager, you help your clients improve their business performance by working with them to resolve as many stakeholder dilemmas as possible. Most of Pentalemma's clients are big companies with offices in many countries around the world. Some of these offices are in unstable or dangerous places. Your client is California-based Speedy Devices Inc. (SD), a global producer of consumer electronics (see Figure 1.1 below). SD now has a problem at one of its suppliers in Africa, in the Democratic Republic of Congo (DRC).

Based in a small town just north of San Francisco, Speedy Devices is the second-largest producer of consumer electronics in the world. Founded in 1973, SD employs nearly 120,000 people all over the world and has an estimated value of $600 billion.

SD has two main product lines; consumer electronics and computer software. Its main hardware products are personal computers and smartphones. Since the 1980s, the company's customer base has continued to grow at a swift pace. SD reports that "Nearly 20% of the human population uses one of our products every day."

SD is a global leader in Corporate Social Responsibility (CSR). It currently has just over 300 retail stores in 15 countries. The company

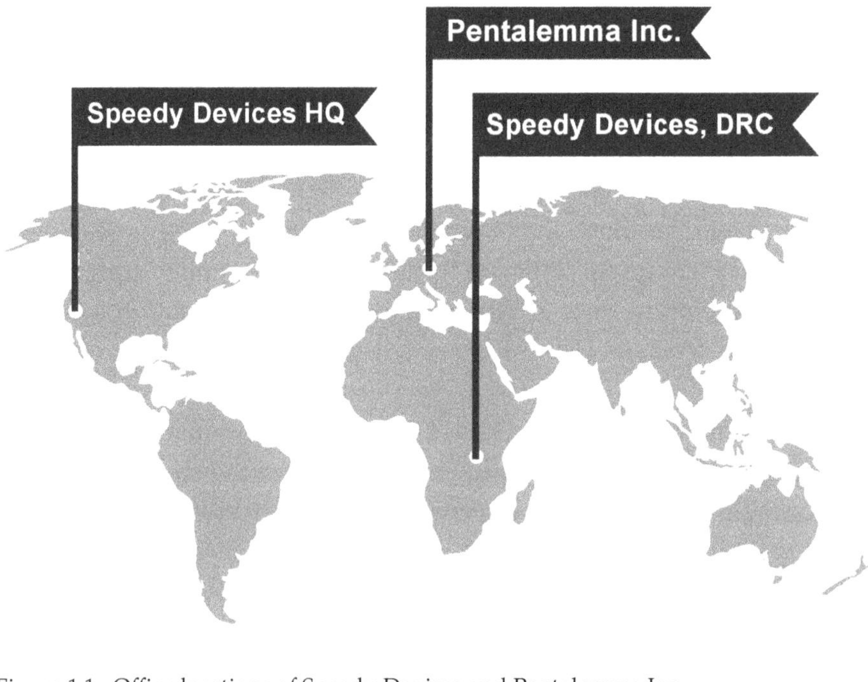

Figure 1.1: Office locations of Speedy Devices and Pentalemma Inc.

also has a strong commitment to society and the environment. SD states in its annual report that its mission is "to manage the triple bottom line so that people, planet and profit are always in balance."

The basic idea of CSR is that a business is part of the community, so the business must act as a member of the community, with rights and responsibilities. The business gains from the community, so it understands that it is important to give something back. It wants to do something good for the community. This means that it should be possible to make a profit and be a responsible corporate citizen at the same time.

However, SD is also one of the world's largest buyers of the rare "conflict mineral" called coltan, which is vital for the assembly of microchip-based electronic devices. Over the last 20 years, more and more people have complained that consumer electronics companies, including SD, are guilty of allowing a range of serious abuses in the coltan business. SD has asked Pentalemma Inc. for advice on the next steps.

Setting

As a Pentalemma consultant, you will focus on South Kivu, a mineral-rich province in the southeast of the Democratic Republic of the Congo. SD secures most of its coltan from South Kivu, where it employs a largely expatriate team of buyers and managers.

Although the province is one of the most unstable regions in the world, the immediate dilemma facing SD concerns child labor. Simply put, the use of young children as labor in appalling conditions is common in the local coltan mining industry. Western criticism of such practices is mounting, with the arrival of an Australian pressure group and recent legislation in the United States of America (USA) requiring firms to use only "clean" coltan. Many of the mining companies themselves are controlled directly or indirectly by a violent militia. It is an environment in which it seems almost impossible to conduct business without being complicit in some way.

To what extent can SD's critical strategic need for coltan be reconciled with this moral and legal challenge? This situation is complicated by ongoing security concerns and the sense that the child labor question may not be as important for local people as western observers imagine.

In meetings with SD managers, you will help them explore and understand:

1. The notion of a complex risk environment, in which diverse and evolving risks overlap and compete for attention.
2. Attitudes to morality and profit, complicated by the notion that ethical issues may be shaped by culture and local circumstances.
3. The potential for cooperation with competitors to address shared risks.

The many issues for discussion include child exploitation, the US Dodd-Frank Act, complex multi-factional security environments, corruption and graft, reputation management and working with Non-Governmental Organizations (NGOs).

Scenario Design

The South Kivu region is arguably the most unstable and dangerous of those covered in the current Pentalemma series. The scenario seeks to demonstrate the complexity of this environment for business, rather than to plunge the readers into the more obvious challenge of dealing with a war. Thus, the war—or its threat—forms a brooding backdrop to the more immediate challenge of child labor. Similarly, there are hints of another dimension of risk; the plight of the lowland gorilla. This is dealt with in a later scenario, but in Chapter One is part of the risk "clutter." Other questions, such as the problem of staff turnover, affect the bottom line and nag at management teams as they wrestle with the central problem.

Two of the options included appear very similar, but are completely different in substance. For while the suggestion for a wide-ranging review is no more than that, and has at its core an ambition to at least scope some options for addressing Kivu's problems, the "business as usual" choice is also couched in nebulous terms of "encouraging dialogue." The intention is to show

how language may obfuscate but also to suggest that a decision to effectively ignore substantial risks will hit the bottom line in due course.

The best option here is rooted in real engagement under local conditions, as understood by the local people. By supporting initiatives and business models that make a start at tackling their issues (primarily security, in this case) a better business environment can begin to emerge. Only then can the thorny question of child labor, considering the opposing perspectives held by local people and the NGO's, begin to be meaningfully addressed. In this way, what seems to be the center-stage risk is, in fact, contingent on broader concerns.

The main learning points here are, therefore, about complexity and risk. In addition, readers should appreciate that morality for business, as anywhere else, appears in shades of grey, depending on the context and viewpoint. To simply insist that SD should have no further involvement in a trade tainted by child exploitation will not solve the problem. In hazardous environments, it can be helpful to cooperate with competitors. The Dodd-Frank Act Section 1502, signed into law in July 2010, requires publicly traded companies to ensure that the raw materials they use to make their products are not tied to the conflict in Congo. Since 2016 the US government has threatened to water down or repeal the act. Even with added uncertainty, the intrusion of such legislation serves as a reminder that political events on the other side of the planet may have a real bearing on businesses trading in Africa. There is a useful analysis of this subject in the "DRC Country Profile" made available by Connie Bernstein.[2] Connie is Pentalemma's main contact at Rapid Response Inc.,[3] the company's worldwide source of intelligence on local developments.

Questions to Consider

The following are some suggested questions for use in your sessions with the SD management team. You will devise your own questions, including those for individual managers involved in specific areas.

1. Should companies try to find ways to work around the requirements of the Dodd-Frank Act, or is genuine compliance a moral imperative?
2. Are there ways to turn security risks into opportunities to improve the situation?
3. If child labor is acceptable within a local setting but not in a company's home country, how should it conduct itself?

Significant risks include:

Social: If the child labor issue in Kivu is left unresolved, then SD could be acting legally, yet unethically.

Economic: Failure to handle the social and environmental concerns in the area could harm SD's reputation, and consequently, its bottom line.

Environmental: The Lowland Gorilla is severely endangered. Unless SD takes measures to protect this rare species, extinction is only a matter of time.

Demographic: Children in Kivu outnumber the adult population. SD needs to accommodate this fact without depending on the exploitation of children.

Political: As the overall security situation in Kivu remains tenuous, SD needs to consider the potential political impact of any decisions.

Defining the Problem

Most of SD's coltan comes from the DRC, one of the most dangerous and demanding business environments in the world. SD's DRC main office is in Kinshasa, the capital. However, there are also field offices in South Kivu, closer to mining operations. There are many risks in the area. If these risks are not managed successfully, the company's global reputation could suffer badly.

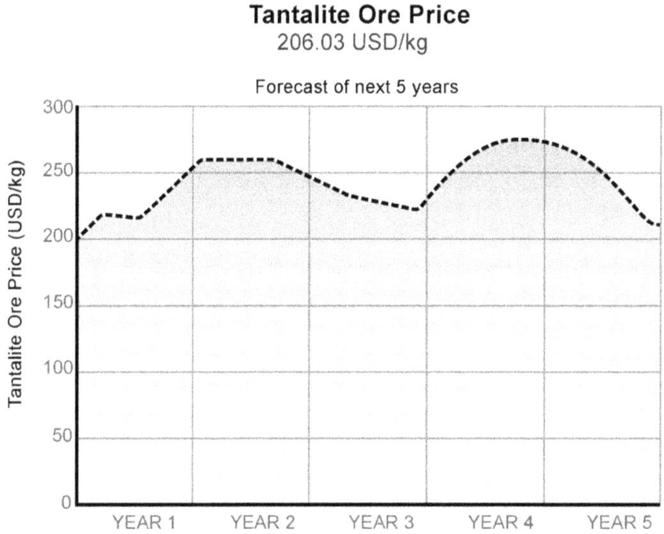

Figure 1.2: Forecasted prices for tantalite ore over the next five years.

SD faces a dilemma. First, coltan is a mineral that every manufacturer in the electronics industry, including SD, must have. Second, there are compelling reasons not to source coltan in this region. You know that children work under harsh conditions to extract the coltan that SD buys. Also, there are various guerrilla groups in the DRC. On your desk is a report from the SD Field Manager in South Kivu Province.

Field Report

SPEEDY DEVICES INC.

From: Karen Adams, Field Manager, Kindu, South Kivu Province
Subject: Field Report—Child Labor in South Kivu Province

The situation here is complicated. We buy coltan through local suppliers. They should not buy coltan from guerrillas, but we cannot control everything they do. The government seems stable. Our staff is not in danger now, but the future is uncertain. In addition, an Australian NGO called "Kids in Crisis" (KIC) has recently opened a small office in Kindu town. They say they will work with the local community, the authorities, suppliers and foreign businesses to fight against child labor in South Kivu.

KIC's goal is "elimination of child labor from the materials used by multinationals." So, I suppose they are here to persuade SD to make sure that children are not used to extract the coltan that SD buys. This would not be easy, because we would have to check where all the coltan comes from. Do we want to cooperate with KIC on this, or not?

There is some background information to think about.

First, in this culture, it is normal for children to work hard from an early age. That is how many families survive here. Second, child labor is not illegal in the DRC, but we need to research the effect of US law. Third, various guerrilla groups are controlling the local coltan industry.

I am meeting with KIC next Tuesday. I need your input before that meeting.

Karen Adams
Field Manager

Applying the Pentalemma Model

It seems that your biggest problem may be the issue of child labor. Of course, SD would never exploit children. Such rules of conduct are written in the company's CSR policy. However, critics such as KIC say indirect exploitation is just as bad as direct exploitation.

This kind of criticism is a huge problem because it damages the company's global image. SD is getting nervous about this, so there is intense pressure on you to propose a solution. The Pentalemma method of problem-solving creates five solutions to the problem. The first solution satisfies one stakeholder, but not the other. For example, it satisfies the shareholders, but not the local communities. The second solution is the opposite of the first solution; it satisfies the local communities, but not the shareholders. The next three solutions are in-between—waiting to decide, compromising between the two main stakeholders or somehow satisfying all stakeholders.

In business, and in the case of SD, this means that you can create five solutions and try to satisfy as many stakeholders as possible. To do this, you need to explore all the possibilities and try to create a win-win situation. By looking at this range of possibilities, the client company can put together a plan, discuss it with everyone and create a win-win for the business as well as its stakeholders.

The head of your planning department, Keith Poulter, sends you the below email with some ideas about how to do this.

Pentalemma Inc.: BOARD BRIEFING PAPER
Client: Speedy Devices Inc.
Subject: Child Labor in South Kivu Province, DRC
Author: Keith Poulter, Head of Planning, Africa and Asia

Staff Turnover

SD needs 23 staff members in the province. It is difficult to find local people with the required skillsets. Ten foreign employees have resigned in South Kivu due to stress-related reasons so far this year. SD should consider better pay and better security for staff in South Kivu.

European Electronics Child Protection Plan

European Electronics (EE) is another multinational corporation (MNC) that buys coltan in South Kivu. Last week, EE announced a new world-wide "Child Protection Plan." SD should monitor this situation closely.

Gorilla Movie

German filmmaker Andreas Brandt plans to create a documentary on the Central African gorillas. Coltan mining is one factor causing their population to decline rapidly. Brandt is famous for his criticism of big business.

DRC Military Situation

Ms. Adams is correct that South Kivu has recently become more stable. The government seems to have control within the area, but rebels could

challenge them again over the next six months. This could increase security costs for SD. It could also put rebels in control of more coltan mining. SD staff may even have to leave the area.
Keith Poulter
Head of Planning, Africa and Asia

The Simple Dilemma

Poulter, Adams and SD representatives initially concur that SD has a choice between prioritizing either profit (you call this Option A) or the environment (Option B). SD managers lean toward Option A as being more realistic than Option B. You begin to analyze these two paths.

Option A—Get Coltan Elsewhere

Getting coltan from South Kivu is becoming very expensive and risky. Sooner or later, war will return to this area. While SD is here, how can it get coltan that has no connection with guerrillas or criminals? It seems impossible to be sure about these things. US law (the Dodd-Frank Act) prohibits dealing in conflict minerals. Coltan is a conflict mineral. Coltan is also connected to child labor and environmental destruction. SD should act according to US law, and we must protect its image. The company may need to acquire coltan from a more peaceful region or country.

Option B—Cooperation

NGOs have two main concerns. The first is that children are being exploited in order to supply coltan to big foreign companies. If military groups are involved, then coltan becomes a conflict mineral. Their second concern is the effect on the environment, especially on endangered gorillas. The solution to both issues is communication, better understanding and cooperation. MNCs should cooperate with KIC and the local authorities to improve the lives of young people. The same applies to the gorilla problem. Businesses can work together with environmental groups and the authorities if they build trust with each other through regular communication. SD must also avoid buying any coltan that provides funding for the local conflict. However, even if all this is done, there remains a major risk that SD's image could be damaged by a media focus on child abuse.

The DRC and Conflict Minerals

With these two options on your desk, you make sure that everyone on the advisory team and the SD management team has read a recent report from Rapid Response Inc. (RRI), one of the world's leading risk management specialists.

COUNTRY PROFILE:

THE DEMOCRATIC REPUBLIC OF THE CONGO (DRC)

by Connie Bernstein

RAPID RESPONSE INC.

In terms of natural resources, the DRC is probably the richest country in the world. However, the country has a long history of war, corruption and poverty. It will be a long time before the DRC has a normal business environment.

Many companies believe the country is too risky for doing business, but other companies do have operations there.

RRI risk rating: SEVERE

DRC Summary

GDP	$17.7 billion
Unemployment	Labor force 35.9 million—unemployment rate not known—population below poverty line: 71%
Currency	Congolese Franc

Ethnicity	Over 200 ethnic groups Bantu people (Mongo, Luba, Congo) and Hamitic Mangbetu-Azande 45%, others 54%, Pigmy (aboriginal people) 1%
Religion	Christian 85% (50% Catholic, 20% Protestant, 10% Kimbanguist, 5% other Christian), Muslim 10% and Animist 5%
Population	75.5 million estimated

Politics and Government

The president is head of state, but there is almost no functioning government in many parts of the country. In the areas under government control, there is a democratic system.

A problem in some areas is that foreign businesses must cooperate with rebel groups. This can be expensive, dangerous and illegal. The Dodd-Frank Act is designed to prevent companies from purchasing minerals directly from guerrilla groups. Although the situation is complicated, some companies have proven that it is possible to do business in the DRC.

Economy and Business

The DRC was once a vibrant, industrialized nation. Today, after years of conflict, it still has enormous potential but suffers from poverty and violence. 90% of the DRC's income comes from selling minerals. There is very little long-term planning. There is almost no manufacturing. Education levels are low, and unemployment levels are high. The DRC is an essential and potentially profitable part of the world, because of its vast mineral riches. However, infrastructure is weak. In most parts of the country, international companies have to provide their own communications systems. The same is often true for electricity and water.

Natural Resources

This part of Africa has the largest deposits of many rare minerals that modern industry requires, including coltan, cobalt and tungsten. The electronics industry cannot operate without these minerals. There are also huge deposits of oil, iron, copper and gold. Water is another resource. The Congo River system has the potential to provide half of Africa's electricity. Agriculture is another major industry. If there is peace, a new ecotourism industry could be developed.

Security

Companies should know that the DRC is a difficult place to work and live because of the hot climate and many health risks. More important, there is frequent fighting between local tribes and guerrilla groups. There are currently 14,000 UN soldiers in the country to keep the peace. The threat of new fighting in the DRC is the main security risk for business.

Conflict Minerals in the DRC

A special update by RAPID RESPONSE INC.

Introduction

The DRC government cannot always control access to rare minerals. Coltan, for example, is often supplied by groups that use the income to buy weapons for conflict or criminal activities. That is the reason why such minerals are called "conflict minerals." The Dodd-Frank Act was passed into US law to discourage trade in conflict minerals. It applies to US companies and companies trading in the USA.

Also, many of the local mining and processing operations have been using child labor. Accordingly, the DRC government has passed laws aimed at ensuring that only "clean" minerals (those not related to war funding, child labor, slavery or other criminal activities) are traded.

There are practical ways of doing clean business, but companies buying rare minerals from the DRC will face ethical, political and legal issues.

It is very difficult to be sure if a mineral comes from a "clean" source, so many MNCs have stopped buying DRC minerals or are considering such a retreat.

Media Coverage

During your SD assignment, it is essential to keep track of media attention on the company and the environment in which it operates. You can't help but notice a newspaper story under the title "MNCs Plan to Pull Out." The article continues:

> Several of the largest electronics companies are thinking about leaving South Kivu province in the Democratic Republic of the Congo. They say that the main reason is the safety risk. South Kivu is a center for trading in coltan. Critics say that there is child slavery in the coltan mines. There are also reports of environmental destruction. If the MNCs stop buying coltan, will the children and the wildlife of South Kivu province have a brighter future?

When this story is brought to the attention of SD headquarters in California, the CEO is angry.

"If nobody invests in that region," he says, "the children have no future at all! Unemployment and poverty will only increase." He calls Option B the "business as usual" approach that really ignores the child labor problem. He immediately asks you to help build a more creative range of options.

Brainstorming

Several intense meetings follow, first involving your Pentalemma colleagues, then including SD managers. Detailed discussions cover possible strategies to satisfy SD's interests, improve the situation for children and protect the environment. Who are SD's stakeholders? Does SD have positive relations with them? What are the ethical and legal issues?

Your team develops three new options for SD to consider.

Option C—Review

We have a complicated and risky situation, so we should start with a careful study. Coltan is an important mineral for the economy, but the child labor issue is a problem. It is always difficult to know what is really happening at the mines.

Everybody—NGOs, the local authorities and global companies—wants to cooperate on these problems. But they need objective and reliable information. A comprehensive study, with information from all parties, is the only way to address this. This should be our first step. A study may not eliminate child slavery, but the world will see that we are trying to do the right thing.

Option D—Preserve and Protect

Ending child labor and protecting the environment should be long-term goals. The situation can be improved by doing the following:

1. Provide primary school education, using resources from NGOs and MNCs operating in the area;
2. Improve health and safety at coltan mines and
3. Establish wildlife reserves in South Kivu.

These initiatives could be the foundation for a more sustainable coltan industry in the region. Success would require continued peace in the region. Considering all these points, this is the right thing to do.

Option E—Village Cooperatives

Without solving the security issue, long-term progress will be difficult. That is why it is important for the MNCs and NGOs to support political stability. Step one is for the MNCs and NGOs to cooperate with the local authorities to build a government and police force that people can trust. This is one way to reduce guerrilla fighting and criminal activity.

Then SD can design an economic model that serves local people, delivers the coltan required for business and reduce environmental damage. Village-based cooperatives would manage the local extraction of coltan and sell directly to the MNCs. MNCs and NGOs would invest in improved extraction techniques, education and training.

Searching for a Win-Win

With five options to consider, you arrange a conference call that includes SD headquarters in California, the South Kivu team and your team in

Kinshasa. The advantages and disadvantages of each option are vigorously discussed.

Option A—Leave South Kivu

The risk of working in South Kivu is too high because of the environmental and child labor issues, and because violence will probably continue in the region. Moving to a new location would make employees happier because they would not have to worry about the company being criticized all the time.

Some feel that changing locations may reduce shareholder confidence in SD's ability to maintain a reliable supply of coltan. Also, if other MNCs leave the area, the local economy will surely suffer.

Option B—Active Stakeholder Engagement

Some participants argue that improving communication with its stakeholders is the only action SD should take. They are especially worried about the effect of media stories about the child labor situation, and Andreas Brandt's gorilla movie. The media reaction will undoubtedly be negative if SD fails to do anything concrete. Its reputation and share price will certainly suffer as a result.

Option C—A Coltan Industry Review

SD does not have enough reliable information about its suppliers. Rather than hastily taking action, SD should cooperate with NGOs and government authorities to do a detailed study of the situation. SD could also cooperate with other MNCs on this. The best way to protect the company's reputation would be to improve the status of all MNCs in the region.

This would also be a better way to improve staff morale. Critics of this approach complain that merely studying the situation avoids solving the problems.

Option D—Preserve and Protect

SD should work hard to provide education for children and improve health and safety conditions at the mines. Consequent good PR and better staff morale would make this a good investment. SD should create a wildlife reserve and restore areas damaged by coltan mining.

One problem with this suggestion is, of course, the cost. Another concern is whether there will be sufficient coltan production if children leave the mines to go to school. Finally, it is possible that the militias would react with violence if SD took away their young workers.

Option E—Village Cooperatives

SD should engage locally with all parties. It can make a long-term investment in the political, economic and environmental future of South Kivu. The only way to create a foundation for peace in the region is to have a stable government and public confidence that the economy will improve. This would help increase supplies of "clean" coltan. It would also solve the morale problem, reduce the risk and cost of security—which is enormous—and make SD's business far more profitable.

Conclusion

Pentalemma challenges you to develop solutions that benefit as many stakeholders as possible. To do this, you have to understand the different perspectives of stakeholders and help them to develop sustainable agreements. SD has some stakeholders who like pure philanthropy. They feel it is more important in this situation, and they believe this is an ethical way of doing business. You must help them to think carefully about what pure philanthropy means for SD's business.

There are other stakeholders who think SD's profitability is the highest priority. They see philanthropy as an unnecessary expense. You should make them see the value of investing in relationships for the future.

Both of these views, the philanthropy view, and the profit view, have a narrow focus that fails to consider other possibilities. You can help managers develop a broader view. Your challenge is to support SD so that the management team can make the best decision for themselves and their stakeholders. There are five options on the table—how will you advise your client?

At this point, let us fast-forward to consider five hypothetical sets of consequences.

Media Reaction

Option A—Leave South Kivu—"Perfect Storm" Sees MNCs Pull Out of Mineral Province

Several of the world's largest electronics companies have announced decisions to withdraw from South Kivu province in the Democratic Republic of the Congo, citing safety concerns as the main reason. Kivu had been the hub for trading in coltan—a rare mineral vital in the fabrication of computers and mobile phones. Frequently an active war zone, more recently the region has seen a decline in local tensions. Yet relative peace brought closer scrutiny of the trade—with criticism alleging routine child slavery in the coltan mines. Added to that reports of wildlife habitat destruction and the continuing security threat and, according to Karen Adams of Speedy Devices Inc., you have "an untenable situation—a perfect storm." It seems that international lobbying can work. The children and wildlife of Kivu province now have a brighter future. As for the multinationals—might some of them be listening more closely these days?

Option B—Active Stakeholder Engagement—Cooperation Promised in South Kivu Province

In what may prove a landmark deal for big business in the DRC's South Kivu province, local companies and NGOs will cooperate to eliminate child slavery and reduce environmental destruction. Children as young as seven are reported to be working in brutal conditions while foreign investors turn a blind eye. Meanwhile, populations of lowland gorillas have collapsed in the face of uncontrolled development. A joint press statement proclaimed: "We believe this settlement will mark a step-change for the children of the area and the natural environment." Meanwhile, insiders warn that the real threat remains as it has been for years: more war.

Option C—A Coltan Industry Review—No Change in Sight for Kivu Kids

Life in the coltan fields is not easy if you are a 7-year-old working a 12-hour day. Until this week, hope had seemed at hand in the form of rising international criticism. Children's advocacy groups like Kids in Crisis had been calling for a complete cessation of child labor in the region, yet a deal announced yesterday means the appeals have fallen on deaf ears. Instead, the NGOs will work with industry and the government in a "wide-ranging review of the coltan economy." Nobody could say just how long that would take, which has led some to suspect that foreign businesses may be intentionally stalling. Meanwhile, children remain exposed to extreme risks, unspeakable abuse and often death, while the area's fragile ecosystem continues to deteriorate. Change may come to South Kivu, but don't hold your breath.

Option D—Preserve and Protect—"Pragmatic Plan" Brings Hope to South Kivu

South Kivu province, DRC: the epicenter of the blood-soaked trade in coltan, a vital component of so many electronics. There is a good chance that your communication device is partially responsible for child labor, natural habitat destruction, frequent violent clashes and multinational companies (MNCs) jockeying for position. But in a deal brokered by the DRC Minister for Industry, NGOs, MNCs and the community have set out what they describe as a pragmatic action plan. The plan, though light on detail at this stage, includes education for younger children, wildlife havens for endangered species and improved health and safety. Can putting this plan into action help to turn this apocalyptic mess around? The world might have hoped for more from the

Ministry conference, but these modest broadly agreed measures might just prove to be the right way forward.

Option E—Village Cooperatives—A Game-Changing Plan for South Kivu

South Kivu in the DRC had become a textbook case of "bad" multinationals destroying communities and environments, all for the extraction of the coltan we need for our phones and a wide range of other devices. So, the announcement on Sunday of a truly radical local plan grabbed the attention of development agencies and businesses alike. NGOs and firms in the area have signed up to improved governance, better security and a cooperative-based extraction economy. It is certainly a leap of faith on the part of multinationals who have been accused of complicity in child labor and environmental destruction. Will the South Kivu experiment weather the storm and provide a template for the developing world? Market results so far: coltan futures up 10% and shares in electronics firms slipping. Watch this space.

Initial Impact on Reputation

Option A—Leave South Kivu

The decision to shift coltan production from South Kivu initially received a lot of coverage. International media have pitched the move as a watershed for multinational business, suggesting that it puts down a marker about both child protection and the need for environmental protection. Only the Financial Times ran a line about dislocation and increased coltan prices; but over time, this angle took hold and SD was accused of betraying shareholders. Locally, there is concern that the baby may have been thrown out with the bathwater. "What now for Kivu's economy?" ran one story. There seems to be little celebration of the child welfare dimension, although some recognition of the potential for future wildlife tourism.

Option B—Active Stakeholder Engagement

Press coverage of the working protocol recently agreed in South Kivu has been mixed. The mainstream media have given the plan a cautious welcome, caveating most reports with a reminder that the overall security situation in Kivu remains tenuous. This may not play well with expatriate staff working on-site. Amongst some of the more "radical" elements of the press—mostly newspapers so far—there have been a few more critical pieces. The chief concerns here seem to be that little concrete action is proposed in connection either with child labor or wildlife protection. If this proves to be the case, then the companies buying coltan in South Kivu may suffer reputational damage. Already, "un-named sources" from within KIC have been briefing against the protocol in the Australian press.

Option C—A Coltan Industry Review

The launch of a long-term review into the South Kivu coltan industry has not generated much media coverage. There is a sense that this will only be "news" when the review finally concludes—or if, in the meantime, war returns to the region. The joint press conference convened by CEOs, NGOs and local authorities to make the announcement was poorly attended. Stories were relegated to the tail-end of bulletins and inside pages. Interestingly, for a few days, Twitter was alive with rumors of a fudge, "long-arming the problem" as one contributor put it. The mini-slide in electronics shares at the weekend was widely linked to the review. Bottom line: the press seems to be suggesting continued uncertainty breeds risk.

Option D—Preserve and Protect

The media have been universally supportive of the reform plan for South Kivu announced at the weekend. Given positive briefings from both the NGOs and CEOs, this is hardly surprising. There is a recognition that this may be "the best that can be done under the circumstances."

Option E—Village Cooperatives

It is being called the "Kivu Experiment" and hailed as a new model for development and security in much of the liberal western media. The deal struck last week to redesign the South Kivu government and the economy has certainly raised eyebrows. Local coverage has also been glowing, with few daring to speculate about failure. The degree of investment required from all parties is significant—hard cash, and above all, cooperation. But they seem to think it worth it. For once, there seems genuine unanimity between NGOs, CEOs and the authorities in the DRC: This is an experiment worth pursuing.

General Impact on Firm Value

Option A—Leave South Kivu

COMPANY REPUTATION: Although SD's decision to pull out of South Kivu initially garnered substantial support in the press, the financial heavyweights overshadowed this with more negative analyses in the weeks that followed. The notion began to circulate that SD had cut and run—letting its shareholders down. Shares tumbled and as the financial results for the year appeared shaky, so did the firm's reputation for competence. Initial estimates of cost are negative $6,000,000.

ON-SITE PRODUCTIVITY: Production fell as SD could not fully implement the shift to locations in more stable regions. Productivity would certainly have been boosted by new technology in the form of a portable dredging and refining plant, but delays were encountered. Even so, trade remained robust, and the estimated net value of productivity in the region is $21,000,000.

RISK MANAGEMENT: SD reduced its risk exposure by moving to more secure areas. However, the availability of government subsidies became very uncertain, and a commercially viable new technology impacted poorly on the company. Deployment of the new dredging and refining technology has been delayed until next year. The efficiencies of such technologies help to reduce risk. Risk management benefits amount to no change in net value.

SAVINGS IN SUPPLY COSTS: While the company operated more production facilities, its shipping costs also increased. But SD's finances remained unaffected as the company achieved economies of scale on coltan exports. However, technology costs to extract purer grade coltan—demand for it is increasing—did eat up savings. Besides, violence made extraction more difficult. There were no estimated savings in the supply costs.

STAFF MOTIVATION: Staff morale improved with the relocation decision. People prefer to work for a company that is not in the middle of a media firestorm for exploiting child labor and damaging the environment. Staff motivation resulted in an estimated gain of $2,000,000.

Option B—Active Stakeholder Engagement—Let the NGOs Do Their Job While We Do Ours

COMPANY REPUTATION: The issue of child exploitation had a significant impact on SD's reputation. The reputation further declined as the company became increasingly visible at the militia-run sites. This was noticed particularly by reporters in the area and German documentary filmmaker, Andreas Brandt. As the company's reputation suffered, its stock price also plummeted. The residual value of SD's reputation has consequently dropped to $3,000,000. A serious attempt to engage with what is an important issue across the planet—child labor—might have salvaged SD's reputation.

ON-SITE PRODUCTIVITY: Production remained relatively stable over the past 12 months as the company decided not to increase production at the existing sites. Also, expanding elsewhere in the DRC was out of the question. By default, in the absence of an up-front one-off expansion cost, the productivity benefit is estimated to be $15,000,000.

RISK MANAGEMENT: The security situation became precarious in the region. Costs increased and losses shot up while, for a brief period, warring factions halted coltan extraction. Employees were evacuated as the situation deteriorated. The overall risk management impact is estimated to be *negative* $1,000,000.

SAVINGS IN SUPPLY COSTS: Supply costs mirrored production in that there was no noticeable fluctuation in either case. The volume of production remained unchanged. The estimated impact on savings in supply costs was $5,000,000.

STAFF MOTIVATION: The low morale of staff despite glowing reports on the company performance is a problem SD needs to address. The low morale stems from the fact that the expats are the favored ones who hold key posts, to the chagrin of the locals. Turnover is still almost three times higher in the DRC region than in any other SD operating area. The value of staff motivation is estimated to be $2,000,000.

Option C—A Wide-ranging Coltan Industry Review

COMPANY REPUTATION: A comprehensive study showed that SD was concerned about its source of labor. It was also concerned about the credibility of previous studies. Partnering with the NGOs, government authorities and competitors helped the company save its reputation from allegations of child abuse and causing environmental damage through coltan mining. Current estimates put the value added to the company's reputation at $10,000,000.

ON-SITE PRODUCTIVITY: Sales remained steady, and there was no increased production. Media attention on the review initiative is not sufficiently negative to impact on the company or its partnerships. Coltan output remains steady for the time being and, therefore, productivity benefits will stay at $15,000,000.

RISK MANAGEMENT: The risk remains high for further conflict if the militias turn on the companies that are recruiting people within the DRC. However, the high visibility of the region among stakeholders, such as the NGOs and the UN, has helped to prevent any further violence to date. The estimated benefits of lower risk management costs are $7,500,000.

SAVINGS IN SUPPLY COSTS: As sales remained steady, so did supply costs. With no increase in production during the past year, supply costs savings were about $7,500,000.

STAFF MOTIVATION: SD's proactive attempt to resolve the child labor issue encouraged employees. They are now less likely to leave their employer. As SD partnered with well-respected organizations, staff morale improved. The estimated benefit works out at $7,500,000.

Option D—Preserve and Protect

COMPANY REPUTATION: SD lifted its reputation by working hard on the social front. Education, health and safety improvements earned the company kudos, although child labor is yet to be eliminated. Creating a wildlife preserve and reclaiming land from old coltan mine sites also did the company and its image the world of good. The estimated benefit to the company's reputation is $35,000,000.

ON-SITE PRODUCTIVITY: As child workers left work to join educational programs offered by the NGOs and SD itself, production could not continue at its usual levels. The negative productivity impact is an estimated $10,000,000.

RISK MANAGEMENT: Security risk remained high. Militias didn't like SD's one-upmanship as the company lured people away from these bandits and provided them with jobs and basic needs—something that even the DRC government could not do. However, the long-term benefits of a better-educated and healthier workforce will mean greater productivity and social stability. Risk management will benefit to the extent of $10,000,000 under this option.

SAVINGS IN SUPPLY COSTS: Although production was down, fixed transport costs meant that savings in supply costs were lower than expected. The overall benefit for supply cost savings was $5,000,000.

STAFF MOTIVATION: Staff morale got a boost from this new initiative. Employees are aligning themselves with a company that cares about the needs of its workers and not just the bottom line. Although security risk demands a constant level of heightened awareness, it can be expected that the benefits will exceed the costs related to morale. Overall, staff motivation benefited $15,000,000.

Option E—Village Cooperatives and Improved Governance—A Radical Plan for South Kivu

COMPANY REPUTATION: Investment in the political, financial and environmental future of South Kivu did wonders for the reputation of SD. The village cooperative scheme took power out of the hands of the militia and put it back to where it belongs, in a stable local government. Changing the economic model seemed charitable for an MNC like SD. The non-economic upside was huge, but there is still an economic benefit to be had, and that is valued at $15,000,000.

ON-SITE PRODUCTIVITY: Other forward-thinking MNCs soon moved to share in "fair trade" coltan. As stability improved, costs related to security (including staff turnover and equipment maintenance) were shorn. The price of cutthroat competition has also been sharply reduced. The productivity benefit equated to $40,000,000 last year.

RISK MANAGEMENT: In bypassing the power of the militias and restoring faith and credibility to the government, the political risk was mitigated under more trusted police and military forces. This benefit is valued at $15,000,000.

SAVINGS IN SUPPLY COSTS: Economies of scale were reached as increased production led to increased discounts with suppliers. The overall value of these savings is estimated to be $10,000,000.

STAFF MOTIVATION: The staff were well motivated last year as they now know their company is doing right by improving both the local economy and the environment. Benefits amount to $25,000,000.

Reflecting on Chapter One

The Scenario in Context

This is the first of three Pentalemma scenarios set in the Democratic Republic of the Congo—one of the most dangerous and unstable business environments in the world. It was designed to introduce that setting with all of its complexity, focusing on the seemingly straightforward tension between the immorality of child labor and the business need for strategic minerals.

At this time the area is at peace, although various guerrilla groups are active in the coltan industry. As well as getting immersed in this highly challenging situation, the purpose was to explore how various risks relate to and influence each other. The child labor issue cannot be considered in isolation. In the context of war and peace, just how important is it? Other themes include the role of competitors and one of Pentalemma's central tensions—that between morality and profit.

Dilemmas and Challenges

A central question for SD in this scenario is how to respond to the growing criticism of the use of child labor. As well as the arrival of a vocal NGO, the Pentalemma Briefing makes it clear that there is now a legal dimension to this question in the form of the US Dodd-Frank Act. So, the company is under both moral and legal pressure to stop being a party to the exploitation of minors. The latter risk is expanded upon in the DRC Country Profile. The DRC is such a unique environment that this profile may well be worth re-reading for succeeding scenarios.

Yet despite the conundrum of child welfare, you and your team of advisors will have been all too aware that this operating environment is extremely volatile. The scenario was designed to encourage teams to examine these broader factors and suggest that they interlink. It may be the case that the SD strategy should give greater weight to the security threat than to the supposed risk to children. Systematic risk analysis is a useful tool in scoping such competing issues.

Karen Adams provided insights into what local people on the ground might actually think about these questions. It is not at all clear to her that child labor is a major local priority. This notion of cultural and moral relativism is important, and it occurs in international business all the time. Managers must often triangulate between moral, business and social priorities, with an eye to securing a broadly satisfactory solution. Alternatively, it may be decided that just one of these criteria—perhaps profit—should be sovereign.

You will also have been aware of other tensions and circumstances in need of consideration: the possibility that the threat to the lowland gorillas would somehow be linked to your client's business; and the double-edged sword represented by SD working with its direct competitors. Picking a way through this tangle in search of an optimum outcome was an extremely difficult challenge.

Notes

1. The roots of the pentalemma perspective can be traced back as far as Aristotle and his square of opposition (Béziau & Payette, 2012). While it only defines four points of opposing propositions, a fifth mid-point of compromise between these can be inferred. The German philosopher, Johann Fichte, introduced the thesis-antithesis-synthesis triad in his 1794/1795 publication, Foundations of the Science of Knowledge (Breazeale, 2013). This triad implies a process of reasoning that develops and then reconciles opposing views. Another German philosopher, Wilhelm Krug, made explicit mention of a pentalemma in 1827 in his dictionary of philosophical terms, defining it as a five-dimensional problem or "five-horned conclusion" (*ein fünf-gehörnter Schluß*).Between the 1950s to the 1970s, Norwegian sociologist and pioneer of peace and conflict studies, Johan Galtung, developed in his lectures and unpublished books a five-point model for analyzing social conflict. These were later published in the Encyclopedia of Violence, Peace and Conflict in 2008 (Kurtz, 2008). I had the opportunity to join many lectures and workshops over a period of several years in the 1990s in which Galtung would demonstrate how to analyze conflicts with the model. At a lecture he once gave in Germany in 1999, he drew the model on a flipchart while telling the students, "I don't understand why more people don't use this." I took note and began experimenting with it from that day. Galtung was the first person to publish analyses of social conflict from a five-point perspective, something he confirmed in a discussion I had with him about this at a conference in 2016 in Japan.In the management literature, Blake

and Mouton (1964) developed the five-point Managerial Grid Model in 1964, differentiating between leadership styles in relation to a concern for people versus concern for production. In 1981, Fisher, Ury and Patton then introduced a similar model for interest-based negotiation in their bestseller *Getting to Yes*.More recently, the model has also been used in the psychology and related coaching literature, as in Thomas Augspurger's 2008 book on constructive leadership. Different versions of the diagram are still used in game theory to depict Nash equilibrium or Pareto optimality. Drawing inspiration from these perspectives, especially Galtung's work on conflict theory, I apply the pentalemma model to the challenge of satisfying the competing needs of corporate stakeholders.

2. Connie Bernstein is a senior consultant at Rapid Response Inc., which she joined in 2011. Connie manages a small team of consultants who undertake detailed research on global "hotspots" and works with clients who need bespoke and intensive business continuity advice. She used every opportunity to keep her skills fresh by maintaining her own client portfolio and working on the frontline whenever she can. Much of her time these days, however, is devoted to building up the RRI Country Profile Service. This new service provides up-to-date summaries of countries around the world and what it means to do business in them.
3. Rapid Response Inc. (RRI) is one of the world's leading business continuity consultancies. The company provides advice on maintaining business in volatile and complex environments. Specialized in crisis management, RRI helps businesses secure their operations under extreme conditions such as natural or manmade disasters. Emphasizing foresight and preparation, the company believes that a well-prepared organization should be able to carry out its business in almost any environment. The client base is diverse, including groups like Pentalemma that advise on broader areas of risk, multinationals, governments and individuals. RRI consultants have been recognized experts in their field since 1985. In the RRI reports, a "high" status does not mean simply "stop": it means stop, review and where appropriate, proceed with caution. Between the extremes of "low" risk and "severe" environments, "guarded," "elevated" and "high" assessments may be regarded as transitional. These show when a country is moving significantly in one direction or another, thereby providing an overview of the security climate at a glance.

Further Reading

Béziau Jean-Yve, & Payette, G. (2012). *The square of opposition: a general framework for cognition*. Bern: Peter Lang.

Blake, R. R., & Mouton, J. S. (1964). *The managerial grid*. Houston, TX: Gulf Pubs.

Breazeale, D. (2013). *Thinking through the Wissenschaftslehre: themes from Fichte's early philosophy*. Oxford: Oxford University Press.

Fisher, R., & Ury, W. (1981). *Getting to yes*. Boston: Mifflin.

Galtung, J. (2008). *50 Years: 100 peace and conflict perspectives*. Oslo: Transcend University Press.

Krug, W. T. (1827). *Allgemeines Handwörterbuch der Philosophischen Wissenschaften*. Leipzig: F.A. Brockhaus.

Kurtz, L. (2008). *Encyclopedia of violence, peace, and conflict*. San Diego: Acad. Press.

Chapter Two

Analyzing Costs and Benefits

Your first year in the DRC was successful. In the end, SD decided to pursue Option E. The media response was very positive:

"People are calling this innovative initiative the 'South Kivu Experiment.' This new model for development and security will help to redesign the South Kivu government and economy. The investment required from all the parties is significant. It will also require close cooperation, but finally it seems that the NGOs, CEOs and the DRC authorities have reached an agreement. Thanks to the creativity and hard work on all sides, there is once again hope on the horizon for DRC."

Unfortunately, however, there is now new criticism that over 90% of the lowland gorilla population has been destroyed because of big businesses like SD. The selling price for one kilo of coltan is now over $200. It is near the surface so that anybody can dig it up without a machine. Thousands of people are coming to the DRC to find and sell it.

For food, these miners eat the gorillas living in the nearby forests. NGOs say, "It's because consumers around the world are buying electronics, and global electronics companies want to make a profit. Soon, the last gorillas will be killed—the lowland gorilla will become extinct!"

A media storm followed the release of a controversial documentary film about the rapid decline of the lowland gorilla, linking this tragedy directly to coltan mining. MNCs in the area are being urged to "do something" as the gorillas' habitat is squeezed and they are killed for bushmeat.

Speedy Devices remains active in the area, and they have asked for Pentalemma's advice regarding a response to these allegations.

The region of South Kivu remains precariously poised between peace and war, with a diverse range of rebel groups jockeying for influence. The coltan mining industry is still in some measure controlled by these lawless bands.

SD is not directly responsible for the gorillas' plight. With local war seemingly imminent and sales of electronics in Europe stagnating, it may be difficult to give high priority to the wellbeing of jungle apes. Yet a well-resourced environmental group has arrived in the region and is putting money into wildlife rangers and other initiatives. Perhaps this will be enough in itself to save the gorillas. It has become clear that SD's international staff care deeply about this issue. Their customers are equally concerned: Environmentalists threaten to take direct action and to encourage a global product ban.

There are several topics that require discussion:

1. The idea that broadening a local economy which is dependent on one industry may mitigate risks for that industry as well as help the community.
2. The manner in which seemingly peripheral issues can quickly become business-critical risks.
3. The appropriateness or otherwise of managers and staff taking a moral or political view and thereby shaping company policy.

The many issues raised in the scenario include the power of modern media, the security of transportation links, endangered species and product boycotts and direct action.

Scenario Design

This scenario was conceived to illustrate the potential for a sudden impact on the business in connection with the plight of Africa's gorillas. Swiftly transmitted and magnified through the lens of global media, be it a major cinema release or merely internet chatter, companies can find themselves on the wrong side of an emotional argument almost overnight. Environmental issues such as this have particular resonance, but a company might be forgiven for thinking this is nothing to do with them. Very quickly, a seemingly marginal issue puts a global company's reputation right on the line.

As the gorilla storm swirls around South Kivu, the threat of war remains very real, and other risks must be carefully assessed before SD makes decisions. Many of these factors feature in the options presented, with implications for metrics such as reputation or staff motivation.

Diverse attitudes to the central environmental question are presented intentionally. The decline of the gorillas means a lot to the "western" staff employed by SD but is by no means as important to the local population in South Kivu. It is for consultants to determine the appropriateness of environmental lobbying from within SD and perhaps to spot an opportunity for both business and the community. The most successful strategy will be to join with NGOs and local people to turn the gorilla "problem" into an asset for SD and a valuable component of the local economy.

Specific learning points arise from these dilemmas. To take an example, it became clear that the opinions of staff can serve to alert senior management to the potential impact of an issue that might otherwise seem irrelevant. The speed with which such a risk can turn into a category red event in our globalized system cannot be overstated. Companies need ever more sophisticated risk-scanning systems. Finally, consultants will appreciate that broadening the economic base of an area in which they operate can bring greater prosperity and stability for all—with knock-on benefits for their own business.

Questions to Consider

1. Was it appropriate or professional for SD's field manager to write that the company had a "clear moral imperative" with respect to gorillas?
2. How can companies best prepare for those seemingly marginal risks which suddenly become critical? What is the role of the media in such transitions?
3. From the point of view of the local people, the gorilla question might be seen as an opportunity emerging from a problem; but to what extent would this be true for SD?

Significant risks in Chapter Two include:

Legal: If the European Union (EU) implements the regulations it is currently examining, SD's coltan would have to be certified as having been sourced from a mine not under de facto control of local guerrillas. This is a difficult standard to meet in an area as volatile as South Kivu.

Demographic: Failure of the coffee crop has sent workers south, seeking employment in the coltan fields and adding to the growing numbers of destitute people in the area.

Environmental: SD and large-volume buyers like SD are what make the coltan industry in South Kivu possible, which in turn has irreparably harmful consequences for wildlife and the natural environment.

Ethical: Although animal welfare is not a priority at the local level, the fact that Africa's gorillas are now prominent in global media puts extra pressure on SD to take an ethical stance and handle the issue responsibly.

Logistical: The railway line from South Kivu into Zambia is prone to disruption, which has negative consequences for SD's profitability. This raises the question of whether it would be worth investing in alternative modes of transportation.

Your task, as always, is to assess the situation and help SD managers make good decisions. You will gather information from Pentalemma and from SD. You will contact experts to get their advice, and you will do your own research. Finally, you will report to the CEO of Pentalemma.

You check your email and find the below report from Karen Adams of SD.

Field Report

SPEEDY DEVICES INC.

From: Karen Adams, Field Manager, Kindu, South Kivu Province
Subject: "Gorilla Gorilla" movie

Andreas Brandt's documentary about the gorillas in South Kivu has been released worldwide. It has attracted much attention. It says that SD and other MNCs have helped to create a coltan industry here and that people in the industry do not care about the natural environment. Gorillas are being killed, and their habitat is being destroyed. Of course, it is not our fault, but the movie will make our customers think we are an irresponsible company.

Many people are worried about the future of gorillas. I think the timing is right to start a project to show our support for local wildlife. After the release of the movie, the Gorilla Gorilla Campaign (GGC) was launched. GGC now has its rangers in South Kivu and works with the DRC government, the World Wildlife Fund (WWF) and other groups. They say that we should change the coltan industry for the South Kivu region completely.

SD's image may be in danger. Journalists are suddenly asking many questions. Also, some staff are worried about what kind of action the NGOs will take.

Besides, the military situation in this area may be getting worse. Should we improve the security of our staff and property? Maybe we should work with some of the other MNCs on this.

I look forward to your advice.

Karen Adams
Field Manager

Is Adams's suggestion correct that "SD should launch a new project" concerning the gorilla situation? SD already has an unambiguous CSR policy.

SD only buys coltan from local mining companies. It does not operate its own mines. It does not cut down trees or kill animals. Do you think the gorillas are a problem that SD should try to solve? Of course, protecting the company's image is partly your responsibility. Now, it seems, people all over the world are wondering whether SD will continue harming gorillas.

Keith Poulter, head of Pentalemma Inc. regional planning department, agrees that the gorilla problem is serious, but regards other issues as more urgent. He sends you the below email.

Pentalemma Inc.: BOARD BRIEFING PAPER
Client: Speedy Devices Inc.
Topic: The real price of South Kivu coltan
Author: Keith Poulter, Head of Planning, Africa and Asia

Coltan is cheap in South Kivu because of the low cost of labor. However, various factors make the cost of obtaining the mineral much higher.

Security, for example, is a crucial issue. Companies pay a great deal to protect their staff and property. At least ten military groups are active in the province. Some of them earn money from coltan mining. This makes coltan a conflict mineral.

The USA already has a law that means companies buying any mineral must make sure that the payments they make are not funding conflict or criminal activity. The EU is doing the same. SD cannot be sure that the South Kivu coltan has not come from mines controlled by military groups or criminals. Creating systems to certify the coltan as "conflict-free" will add a lot to costs.

Another potential cost is damage to the company's brand or reputation. If consumers or investors think that SD is using conflict minerals, it could lower the company's sales or share value.

Finally, in the DRC, companies must often provide their own transportation and telecommunications. For example, the railway line from South Kivu into Zambia was cut by rebels last month. Such things can happen at any time, and delays create costs.

Costs may keep rising, and the company's image could be damaged. SD must decide if it is practical to continue doing business in South Kivu.

Submitted for your consideration.

Keith Poulter

Head of Planning, Africa and Asia

You collect opinions from various experts, including Poulter and SD field manager Karen Adams.

Adams responds as follows, copying SD HQ:

As per your request, I have discussed with various people within our network. We seem to have two main options, giving priority to profit, or giving priority to the environment. I prefer the second option.

Option A—Put Profit First

Gorilla Gorilla Campaign (GGC) rangers are already patrolling gorilla areas. There is not much more business can do to protect gorillas.

SD will be blamed for the shrinking gorilla population either way. Our image will suffer, and the share price will probably drop. However, SD must have coltan. War will likely return to the region within a year, so we should quickly get as much coltan as we can. Meanwhile, we can ask the government to improve security and stability.

This is a dire situation for SD in the DRC, yet there is little that we can do to lower these risks. We do not know how coltan mining affects the environment, so we should concentrate on our own business.

Option B—Major Environmental Contribution

We must accept that, when people think about buying our products, they consider the environment, as well as price and usefulness. GGC is considering a product boycott, including SD products.

It is time for SD and others to actively protect local wildlife. The solution is to meet with GGC and the WWF and develop an agreement about future coltan mining. The question is whether we should offer a one-time "environmental donation" or something more.

It will reduce our profits if we provide long-term support over the years. However, environmentalists want a long-term partnership.

SD should build an image as an environment-friendly company. Coltan from South Kivu should be seen as "gorilla-friendly." Achieving this will cost money, but the investment is necessary for our long-term success.

Karen Adams
Field Manager

Which of these two options would you choose? Each has advantages and disadvantages. In consultation with SD and Poulter, you decide to call in Fernando Rodriguez.[1] Pentalemma Inc. has used his services for many years, and he has a large collection of useful management tools. You arrange a Rodriguez-led workshop for the SD management team on How to do a Cost-Benefit Analysis.

Cost-Benefit Analysis

Cost-Benefit Analysis, Rodriguez explains, is used to estimate the financial results of decisions. We can then compare that decision with other possible decisions. This process can help us to see whether we will get a good return for the resources we expend.

Cost-benefit analysis compares all the costs and benefits of a project. If the benefits are greater than the costs, then the project is a good investment. The costs and benefits include money, time and other resources that have a financial value.

"When there are various options available, calculating the value of each option helps us to choose objectively. We can make rational decisions instead of emotional ones," he says. In the workshop, he explains that rough estimates can be used if the exact numbers are not available. He tells the group to estimate the financial value of intangible things such as goodwill and image.

We can choose from among various levels of precision. For example, we can be very precise and include factors such as interest rates for long-term costs and benefits and numerically calculate the return on investment (ROI). If precision is not possible, we can use a ranking system

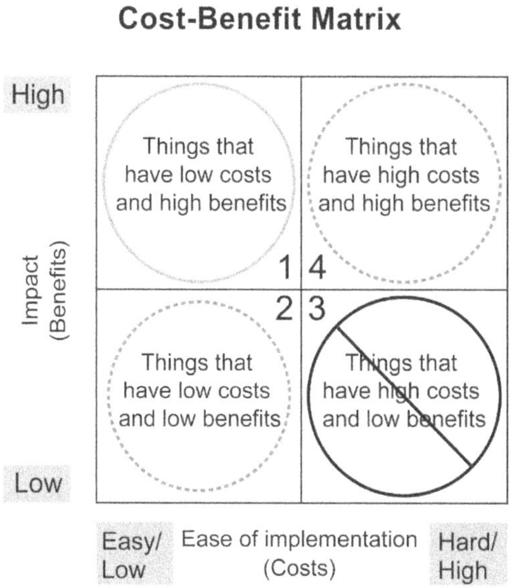

Figure 2.1: Cost benefit matrix based on impact and effort.

instead of calculating everything. Figure 2.1 below shows how to rank the options. Even when complete costs are not available, it is a good idea to estimate the financial results of different choices as best we can.

To create this matrix, workshop participants put ideas on Post-its and then put the Post-its where they think they fit on the matrix. Things that are easy (low costs) and have a high impact (high benefits) are in quadrant 1. The items in quadrants 1 and 2 are the most acceptable. Items in quadrant 4 require more analysis because of the higher costs. The items in quadrant 3 can be ignored.

Now What?

Rodriguez then discusses what kind of data should be collected in order to do cost-benefit analyses for the options available to SD. For example, he asks them how to decide the value of the company's image? Would Option A increase or decrease the value of SD's image? What is the value of staff motivation? How much would the value of staff motivation change?

Rodriguez teaches the group how to do both rough and precise estimates. He suggests that they work with their best financial analysts to find reliable data.

You feel that the management team has learned a lot from Rodriguez. They are now able to use cost-benefit analysis to answer questions related to coltan mining and the gorilla issue. Meanwhile, things are happening fast.

Poulter sends you an online opinion piece titled "Gorillas Caught in the Conflict." The article reads in part:

> There seems to be no hope for the shrinking population of gorillas. Western MNCs such as Speedy Devices did not help. It seems that SD wants the Gorilla Gorilla Campaign rangers to solve the problem by themselves. Many people disagree with SD.

That same day, you see a report on television called Last Hope for Brandt's Gorillas.

> MNCs in South Kivu are making a plan to reduce coltan mining here for the next five years. It seems that Speedy Devices is considering a deal with local authorities and NGOs to establish a large gorilla reserve.

Karen Adams from SD calls you. "Have you seen the media reports?" she asks. "This is now a very difficult situation for us. If we don't take action, the public will think we are irresponsible. But, if we take action to establish a gorilla reserve, it will be costly. What should we do?" She tells you to immediately develop some suggestions for the next steps.

Brainstorming

You can see that neither Option A nor B will solve SD's problem. SD must have coltan in the short-term, but the gorilla issue is escalating very quickly. How can SD continue buying coltan without damaging its reputation?

Using the Pentalemma method, you reconfirm the situation. Who are SD's stakeholders? What do they need? How can you develop more balanced options for SD to consider? After several hours of deep discussions with SD managers and your own colleagues, three new ideas emerge for SD to consider

Option C—Wait for Reliable Data

GGC rangers have reduced the killing of gorillas. The situation may no longer be urgent. SD should take some time to confirm the facts before making a decision. The first step for gorilla survival is for MNCs to support research by UNESCO and the WWF. These studies will be published later this year.

Before these studies are published, there may be other reports that the number of gorillas is declining. In that case, GGC may launch a boycott of SD's products. A boycott would be very costly. The next few months will be a risky period for SD. However, the company cannot take action without the UNESCO/WWF report. It is better to wait than to quickly start a project that may fail because we do not have sufficient information.

Option D—Limited Cooperation

Another possibility is to engage in dialogue with all the stakeholders. SD can continue buying coltan in South Kivu and protect the environment at the same time. The two sides are beginning to see that they have something in common.

Coltan mining is going to continue. The local people and the authorities both want that. However, SD and others cannot say they have no responsibility for the negative results of the coltan business. Also, it is not realistic to allow unfettered mining of coltan.

All sides need an agreement that does not let mining grow too fast, and that provides zones for the safety of wildlife. SD can reach an agreement with the NGO's and the local authorities that allows SD to continue buying coltan for a reasonable price.

Option E—Broaden the Local Economy

SD has an opportunity to meet with the mayor of Kindu, a representative of the government, and GGC. This is a chance to propose that all parties contribute to a large forest reserve for gorillas and other animals. Promoting eco-tourism at the reserve would generate significant income for wildlife conservation. How can this strategy benefit SD? First, the public image of SD would improve. Second, it would enable a more stable supply of clean coltan.

This proposal would help the local economy to grow. It would attract more interest and money from tourists. It would make the MNCs, valuable members of the community. It would also help to stabilize the community and the government.

Searching for a Win-Win

With five options to consider, you arrange a conference call that includes SD headquarters, the South Kivu team and Pentalemma consultants. Opinions vary widely.

Option A—Put Profit First

Many think it is a good idea for SD to get as much coltan as possible before civil war closes all the mines. For these people, gorillas are not the most important problem. Their opinion is that the gorillas are simply in the wrong place at the wrong time. Others argue that this option is risky. They say that SD will be blamed for driving the gorillas to extinction if they ignore the issue. That could damage the company's image for many years. Many SD employees would not be happy if the company showed that it does not care at all. Some of the management team are genuinely concerned about the animals and their environment.

Option B—Major Environmental Contribution

Some participants feel it would be best to make a long-term partnership with GGC. They say this would quickly make SD "gorilla-friendly" and prevent a GGC boycott of SD products. The effect on staff motivation would be positive.

Option C—Wait for Reliable Data

It would be better for SD to postpone the decision until more information is available. Everyone is thinking about gorillas now, but next month they will be thinking about a different problem. SD can wait and see how the situation develops. SD does not need to worry about a GGC boycott.

SD's biggest problem is security, not the environment, according to many employees who don't agree with GGC's objective. They think it would be better for SD to spend money on the safety of personnel.

Option D—Limited Cooperation

SD should agree to buy limited amounts of coltan and contribute to a gorilla reserve. This would become good PR for SD. It would also help to prevent a GGC boycott. However, others say that SD cannot afford to limit its supply of coltan.

Option E—Broaden the Local Economy

The idea of SD and GGC co-branding a wildlife reserve would make SD a leader on the gorilla issue. Employing local people will help to make the region stable and reduce violence. In the long term, this would reduce security costs, and staff morale would improve. Opponents of this option, however, point out that it would be very costly for SD if the plan does not work.

Conclusion

As a Pentalemma consultant, you always have the challenge of developing solutions that benefit all stakeholders. This means that you have to know who the stakeholders are and what they want. It also means that you have to be creative in order to imagine new possibilities that satisfy as many stakeholders as possible. In this second year in the DRC, your core dilemma is profit versus survival of the lowland gorilla.

Once again, let us fast-forward to consider five hypothetical sets of consequences.

Media Reaction

Option A—Put Profit First—"Gorilla Gorilla" Stars Caught in the Crossfire

As rebel groups move south into the DRC South Kivu province, there seems to be little hope for the shrinking population of lowland gorillas. The great apes, immortalized in German film-maker Andreas Brandt's moving documentary, are on their way out. Hopes that western multinational companies (MNCs) such as SD might help the situation have been dashed. Following a recent conference, SD said it was confident that the Gorilla Gorilla Campaign (GGC) rangers would be able to contain the problem. But as SD's coltan mines encroach on gorilla habitat and as the war moves south, few people share that view.

Option B—Major Environmental Contribution—Extraction Deal Brings Last Hope for Brandt's Gorillas

In the face of a threatened product boycott from GGC and other wildlife groups, MNCs in the DRC have agreed to limits on their coltan mining. The rare mineral, an important material in fabricating millions of electronic goods, is scraped out of the ground in the South Kivu area, drawing in itinerant workers and ruining the last of the lowland gorillas' habitat. Andreas Brandt's Oscar-winning documentary put their plight center stage. The deal means that a large sanctuary is to be established, and there will be no more mining expansion in the area for five years.

Option C—Wait for Reliable Data—Andreas Brandt "Loses It," as Gorillas Await UNESCO Study

The German filmmaker reacted with anger yesterday to news that efforts to save the gorillas of the DRC have been put on hold. "This is simply disgraceful," he told a news conference. "The last of the gorillas will be gone before the UN finally makes its report; we do not have time for further delays. I'm calling on GGC and others to boycott Speedy Devices products right now." Karen Adams of SD claimed that Mr. Brandt had over-reacted. "We want to save the gorillas as much as Mr. Brandt does, but we need to understand the problem before we act. SD will look positively on any recommendations emerging from the UNESCO review," Ms. Adams said.

Option D—Limited Cooperation—Development Plan for Brandt's Gorillas

A multilateral development plan is to be put in place in South Kivu, designed to protect local wildlife and allow for controlled mineral extraction to continue. Announcing the plan, SD's Karen Adams described it as a "grown-up deal, to which all parties are fully committed." It seems that the popular GGC has not got everything its activists wanted out of this, but that they will be calling off their boycott. Meanwhile, companies such as SD will cooperate in scaling back the production of coltan. There was a mood of optimism at this conference that hasn't been felt down here for some time. But when one local resident was asked what she thought, she stated that gorillas were beside the point: "Our real worry is rebel guerrillas; not the furry ones."

Option E—Broaden the Local Economy—Gorillas Rock!

That was the message from a conference here in South Kivu this week. Local musician Joshua Nkruma's catchy rock anthem, "Gentle Giants," has been released to publicize and raise money for the new gorilla initiative in this area. The song has gone viral, with millions of hits on YouTube and millions of dollars pouring into the new fund. According to Mr. Nkruma, though, the really big news is the sustainability of the South Kivu plan. "We have the multinationals, the government, and above all, the local residents all signed up to this. It will also give us a new industry for when the minerals run out." The South Kivu eco-park will protect thousands of hectares of virgin bush and safeguard the last of the world's lowland gorillas.

Later Impact on Reputation

Option A—Put Profit First

The "non-deal of the century" was how one left-of-center newspaper in Paris described it. The effective break-up of talks between coltan producers and wildlife groups on Friday, however well camouflaged, has not received much positive coverage. Andreas Brandt, the German filmmaker who made the "Gorilla Gorilla" documentary, did a long interview with CNN in which he pulled no punches vis-a-vis the multinational companies (MNCs) in the Democratic Republic of the Congo (DRC). Their reputations have definitely taken a knock here.

Option B—Major Environmental Contribution

Most of the print and TV media give credit to industry for stepping up to the plate in the struggle to save the lowland gorilla. This was helped by some very supportive copy from Gorilla Gorilla Campaign (GGC), Mr. Brandt himself and others. The financial press (Milan, New York and London) has commented on the likely impact on world coltan costs. In Germany, one daily ran with "Multinationals cave in to environmental lobby." You can never please all of the people.

Option C—Wait for Reliable Data

There has not been much coverage to date, but whatever has been received is quite supportive. There was a "Time-is-running-out-for-the-gorillas" piece in the Washington Post, and Brandt called his own press conference to berate SD, but the overall impact appears to have been limited.

Option D—Limited Cooperation

Cautious optimism would be the best way to summarize the media reaction to the recently announced agreement in South Kivu. There is widespread coverage of the GGC's robust determination to make this work and its rejection of a product boycott. London's Financial Times called this "grown-up environmentalism." There was an interesting piece in Der Spiegel (which previously gave Mr. Brandt huge prominence) from the point of view of local villagers who seem uneasy that future coltan extraction will be restricted.

Option E—Broaden the Local Economy

The widely drawn deal struck in South Kivu last week achieved huge media attention. The release of the pop song online served to snowball the effect so that all eyes were on this remote part of Africa. Enthusiasm and optimism jumped from every page—even the Times of London. It was hard not to share this, given the smiles from SD's normally stern Karen Adams on BBC World. This deal was a media triumph but, doubtless, there will be close scrutiny if it begins to unravel.

General Impact on Firm Value

Option A—Put Profit First—Extract Coltan While That Remains Possible

COMPANY REPUTATION: Speedy Devices will continue to be blamed for the poaching of gorillas for bushmeat as we continue to extract all the coltan we can. Coltan supplies are even more important than any potential backlash from the GGC, WWF or local government. We estimate a reputational value loss of $1,000,000.

PRODUCTIVITY ON-SITE: SD's perceived culpability in the killing of gorillas in the region will have a negative impact on sales and reputation. In contrast, new deposits that have been found in the western part of the DRC promise high returns due to that region's stability. The current strategy implies a productivity benefit of $9,000,000.

RISK MANAGEMENT: This option maximizes risk. It maximizes the environmental, social and ethical risks all at once. The film and GGC both believe that SD has fostered a working environment with complete disregard for the surrounding environment. News of the film also spread fast and increased the risk of a product boycott as well. The irresponsible way in which this option chooses to proceed does not minimize any of these risks. Finally, the constant threat of war in the area keeps all costs high. The cost of risk management is negative $4,000,000.

SAVINGS IN SUPPLY COSTS: There is no saving in supply costs with this option. Already this year militants sabotaged the main supply route. Quickly repaired as it was, the ever-present danger of terrorism can cause supply costs to skyrocket. Although the railway tracks are now better guarded, there is still some risk associated with potential transportation disruptions. Supply savings are estimated at zero for the current year.

STAFF MOTIVATION: Staff motivation will continue to suffer whilst SD maintains this stance. Many employees are unhappy that management is ignoring an issue that is now getting worldwide attention. Staff motivation is valued at only $1,500,000.

Option B—Major Environmental Contribution

COMPANY REPUTATION: Speedy Devices's reputation is now that of a "gorilla friendly" organization as it partnered with the GGC last year and made a pledge to sustain support for the GGC for years to come. GGC's rumored boycott against SD never occurred, which saved our company millions in public relations and various forms of advertising. SD is now seen as the forerunner in relation to environmental concerns in the coltan industry. This has a net value of $5,000,000.

PRODUCTIVITY ON-SITE: Without the strain of constantly having to commit money and personnel to ward off criticism, and with employee unhappiness dissipated, South Kivu operations are running much more smoothly. We estimate that net productivity gains of $7,000,000.

RISK MANAGEMENT: There remains the ever-present risk of militias and violence that seem to be native to the area. By gaining the trust of the GGC however, SD mitigated environmental risks to almost nothing, the net value of which is $3,000,000.

SAVINGS IN SUPPLY COSTS: Supply cost savings were met as targeted last year. As demand for coltan from environmentally friendly producers grows, so too will the economies of scale and purchasing power that SD will secure. Freight prices to the north remained high, however. $1,000,000 net effective value.

STAFF MOTIVATION: What's not to like about working for an environmentally conscious company that has the best interests of all at heart? While local staff may not have an opinion one way or the other, the expats in the country are proud to work for a company that walks the walk. The net value of staff motivation is $750,000.

Option C—Wait for Reliable Data

COMPANY REPUTATION: Speedy Devices's "wait for the UNESCO report" strategy just happened to work, avoiding a boycott against the company's products. In fact, SD can now claim to be "gorilla friendly." The film from last year is no longer garnering attention as people have moved on to the next crisis and largely forgotten about the gorilla population in South Kivu. Investors responded to news of the UNESCO report by boosting the SD share price. Net value in terms of reputation is put at $15,000,000.

PRODUCTIVITY ON-SITE: With no boycott, coltan production could remain steady. The goal now is to consolidate and move forward, using the new environmental studies to beef up SD's eco-friendly image among like-minded stakeholders, including those in South Kivu. The net value in terms of productivity is estimated to be $7,500,000.

RISK MANAGEMENT: Security threats continue to be at the top of everyone's mind when it comes to risk in this region. With the environmental risk thwarted by the promising results of the UNESCO study, the focus has turned to ramping up security at the plants and transport links. The prevention of crime is difficult to assess, but we would estimate a net value of $8,000,000.

SAVINGS IN SUPPLY COSTS: Speedy Devices is on pace to meet the financial targets it laid out earlier in the year. The fact that there was no boycott allowed us to keep our supplies moving through the supply chain. Therefore, there is no change, in the net value of supply costs savings.

STAFF MOTIVATION: The fear of living around militias and warring factions is still constant, but employees are rewarded well. The staff seem reasonably content with SD's efforts to address the issues raised by Brandt's film, with most resenting his allegations of delay and taking the side of their employer. However, some concerns have been raised directly with management. All told, staff motivation is assessed at $8,500,000 in net value.

Option D—Limited Cooperation—A Balanced Settlement

COMPANY REPUTATION: SD and the GGC reached an agreement whereby SD would restrict future development of coltan extraction and provide for a wildlife sanctuary for the gorillas. This was well-received by all parties. The limited growth and profitability that this implied received a mixed reception by the markets and within the business community. The net value of SD's reputation for this reporting period is $9,000,000.

PRODUCTIVITY ON-SITE: More controlled extraction of coltan from the South Kivu region, with anticipated steady demand as well as supply over the next few years means that managers have been able to focus on efficiency throughout the South Kivu operations. In terms of productivity, we estimate a net benefit of $11,000,000.

RISK MANAGEMENT: Working with the GGC, meeting their needs and finding a compromise mitigated the risk of a direct product boycott. It also kept the locals and some of the displaced migrant workers employed, meaning they were less likely to join up with the militia forces that occupy the surrounding areas. The net value for risk management added was $7,500,000.

SAVINGS IN SUPPLY COSTS: Supply cost savings were earned this year as predicted. There was no interdiction to any of our supply routes and demand for our eco-friendly product remained level across the world. SD was able to meet its goals for the year in this area. These equated to a net value in savings of $10,000,000.

STAFF MOTIVATION: Motivation among the local workforce remained high as their jobs were not interrupted by a work stoppage, and they were able to continue to earn a living wage. SD's management team on the ground also had high motivation, with the sense that a reasonable deal for the medium term had been struck here. The prominent role played by Karen Adams in this affair secured significant staff buy-in. This resulted in an additional net value of $8,500,000.

Option E—Broaden the Local Economy—A Radical Plan with an Eye to the Future

COMPANY REPUTATION: The co-branding of the wildlife reserve by GGC and SD greatly improved the company's reputation. By being a forerunner on the issue and donating some of SD's resources, the company has become a benchmark for other coltan producers in the area. Net value of this benefit is $14,000,000.

PRODUCTIVITY ON-SITE: Speedy Devices was able to increase production from the previous year under the terms of this new deal. Word of our environmentalism prompted other companies to sign exclusive contracts with SD. By taking a cooperative and creative approach to the gorilla issue, SD is in a much stronger position than it might have been. The estimated effect on productivity is $22,000,000.

RISK MANAGEMENT: The main concern is still the bushmeat issue that blights the area. With local and international support now driving the PR train, the incidence of poaching should begin to subside. Our other main risk, that of security, is receding with the adoption of a plan which offers a genuinely sustainable economic and environmental future for the region. It is this more than anything which will undermine support for rebel groups.

Our risk management has an estimated net value of $13,000,000 for this year.

SAVINGS IN SUPPLY COSTS: Our savings in supply costs met our estimate from the prior year's forecasting. While there is nothing to this approach that speaks directly to supply cost savings, we should be able to develop more economies of scale and purchasing power from our suppliers and with our distributors. The net value of this approach yielded $8,000,000.

STAFF MOTIVATION: The staff are motivated in that they are part of an environmentally conscious organization that is working with the grain of local society. Local employees are reassured that their jobs will continue and optimistic about the prospects of an improved economy from eco-tourism. Altogether, staff motivation accounts for an added value of $11,000,000.

Reflecting on Chapter Two

The Scenario in Context

Moving on from the child labor question—which is assumed to have been dealt with, at least for now, this scenario again takes place against the real possibility of civil war in South Kivu province, DRC. Its central theme, however, is environmental, with the accusation that multinational companies such as SD have been complicit in the population decline of the lowland gorilla.

The challenge is about how or whether SD should respond to this. Is it anything whatsoever to do with their legitimate business? Unlike the child labor dilemma, there is no obvious direct link between SD and the plight of the apes. It is striking how fast an issue like this can climb up the public agenda, as well as the role of their staff in such debates, and the value of broadening the local economy.

Dilemmas and Challenges

For Pentalemma and the senior management at SD, an early consideration will, therefore, be whether or not to engage with this issue at all.

This is complicated by the fact that Karen Adams makes it clear that the question of gorilla survival is important to her personally, as well as to many in her team. By the same token, we learn that the degree of local commitment to preserving this species is questionable—after all, "bushmeat" is an aspect of the culture. How will SD's actions play into this nexus, and what might the feedback be in terms of reputation, productivity and other factors? Consultants would be wrong to view the situation solely in terms of the gorillas and their supporters. As ever in Pentalemma, and indeed in real-world management, the picture is more complicated.

This remains at heart an environmental issue, albeit one with perhaps only a tenuous link to the activities of the company. In many real situations, such links will be much clearer—as with industrial pollution, for example. In all cases, though, the fundamental question arises as to what degree of responsibility a company might have toward its environment. Management requires familiarity with environmentalism as a popular movement as well as with the academic literature on relevant sub-themes. Patricia Hoenig, one of six mentors at Pentalemma and formerly a Pentalemma Director, has shared some of her recent thinking on the topic at a recent board meeting.[2] She is critical of multinationals and their activities. One of the authors she cites is Slavoj Zizek, who poses several tough questions about what it means for companies to be socially responsible.[3]

The backdrop in this scenario remains one of poverty, tribalism and a multiplicity of rebel groups. Almost the entire South Kivu economy relies on the coltan industry, with much of that income diverted to fund the war. The influx of itinerant workers mentioned by Keith Poulter can only add to the tension. Thrown into this mix is the Gorilla Gorilla Campaign, which is heavily funded and has sophisticated media skills. There is, therefore, a sense in which "fixing" the gorilla reputational risk simply plugs the latest hole in the dike. The point is, therefore, to look for courses of action that might serve to address more than one of these problems.

Notes

1. For the past 15 years, Fernando Rodriguez has trained management professionals in the latest tools and techniques, building for himself a reputation for helping companies develop lasting improvements. His approach begins with building a consensus on the definition of the problem to be solved. From there, he employs several methods of gathering ideas, such as brainstorming and mind-mapping, and then harvesting potential solutions by using an affinity diagram or similar tool. In short, Fernando works with managers to clarify their goals, and agree on the changes they need to make to achieve them. In his words, "I simply help managers become dynamic agents of change themselves."
2. Patricia is a senior advisor to Pentalemma. She began her career over 20 years ago, working for a major agrochemical company in the Midwest of the USA. Her first degree was in Economics, and her role was initially analytical, modeling and planning global production trends in the agrochemical sector. From there, she quickly entered mainstream corporate management in the planning department of the Head Office, completing an MBA on the side. She joined Pentalemma early in her career, working with clients worldwide, advising on doing business in complex environments.Over the years, Patricia began to question the ethical dimension of her work and whether the business focus was the right one for her. The plight of the indigenous peoples in some Latin American countries, for example, and her company's role in that dynamic, began to disturb her. She experienced a crisis of conscience. She had convinced herself that big business was a central part of the problem, and was essentially irredeemable. She resigned in and took six months out, traveling and reading broadly. Still frustrated with the corporate world, she found a job with Farming Without Harming (FWH), a non-governmental organization. Her work was advisory, Washington-based, and focusing on creative solutions to the challenges of modern agriculture. She began a PhD, researching the impact of multinational corporations on agricultural economies in East Africa, combining this with part-time work at FWH.In spite of her misgivings about the corporate sector, she is more convinced than ever that the time for change is now. She is still haunted by the things she has seen in the field; children aged ten working a 12-hour day for less than a dollar—for a well-known MNC; and a yellow, dead river—poisoned by another household name. Her list of concerns is long. Patricia donates part of her earnings to two grassroots movements for social justice.
3. Patricia finds in Zizek's work themes that may be relevant for doing business in crisis zones. The minutes from a recent board meeting include Patricia's comment that "Zizek, and indeed plain common sense, tell us that there is something wrong with our current system of economic production and distribution. Even the big growing economies like India retain huge numbers of people living in relative

poverty and with no social safety net. Matters are much worse in poorer countries, many of them riven by conflict." She adds that efforts to protect the environment are failing and that we are not achieving the carbon targets set out in the Kyoto Protocol. "Eating wild fish from the sea is something our grandchildren may never experience," she warns. In her most recent policy recommendation to Pentalemma, she elaborated on this as follows:"Some might say that Zizek is wrong to claim that capitalism is over, but he certainly has a point. This raises the question of what big business can do to build a sustainable future. If it does nothing, he will surely be proved right. Many business leaders have told me that these are matters for governments to approach through regulation and treaties. But, just because we make top-down progress doesn't mean we can't also make bottom-up progress.Another of Zizek's observations is that people and the planet should have priority over profit. Agree or disagree, Pentalemma has always tried to advise its clients to consider the impact of their operations on all their stakeholders. Companies are not separate from the communities in which they operate. If they want to continue to benefit from the marketplaces and consent of these communities, they have to be committed to the things those people care about. We have to remember, too, that modern communications make it impossible for companies to escape the public eye. If a clothes factory in Dhaka employs children in slave conditions, that news will get out. If indigenous people are thrown off their land to make way for mining, that news will get out too.If capitalism is to prove Zizek wrong, it's clear to me that Pentalemma should do what it can to ensure its clients transform themselves into agencies for positive change. Tinkering with a little philanthropic work, while continuing with a fundamentally exploitative model, is not good enough. The way forward is to think creatively and act responsibly!"

Further Reading

Klein, N. (2007). *The shock doctrine*. London: Penguin Books Ltd.
Žižek Slavoj. (2018). *First as tragedy, then as farce*. London: Verso.
Žižek Slavoj. (2002). *Did somebody say totalitarianism?* London: Verso.
Žižek Slavoj. (2012). *The year of dreaming dangerously*. London: Verso.

Chapter Three

Brainstorming Solutions

You are now in your third and final year of the DRC project. You and your team are in the middle of a post-conflict zone. Local fighting escalated at the end of your second year but has recently stopped. Despite the bloodshed and chaos, and that fighting continues nearby, MNCs such as SD have managed to maintain their activities. Several guerrilla groups created an alliance and defeated the central government troops. They have established a new territory, and new government called the "People's Republic of South Kivu" (PRSK). Fighting could resume at any time in the PRSK as well.

PRSK may nationalize the coltan industry. China has become increasingly influential in the area and representatives from Chinese companies have offered to buy all coltan production directly from the new government. At the same time, some villages have said that they could create a village cooperative and sell the coltan to foreign companies. SD urgently needs to decide whether to leave now or try to do business with the new regime. It is far from clear that it has legitimacy or will last; but for now, it is the "only show in town" and if SD is to continue to operate here, some form of accommodation with it must be found.

The new government is offering a modicum of security but presents itself in the radical language that is often more associated with the late 20th century.

Inflation is extremely high, and staff morale is still recovering.

The UN Security Council has met to discuss the situation. If the fighting returns, French and Nigerian troops have agreed to help foreign staff to escape the country. Meanwhile, the Christian church, one of the few trusted and trustworthy local agencies, is trying to organize peace initiatives.

In your recent discussions with SD Field Manager Karen Adams and Pentalemma's regional Head of Planning Keith Poulter, you agreed that business, government and environmental interests are all important. You all feel that there must be more cooperation with local people. This would help SD to get a regular supply of coltan and prevent damage to SD's global reputation.

Once again, SD faces a dilemma: continue doing business in South Kivu, or leave the DRC and buy coltan from another country. From your research, it seems SD must stay because the electronics industry cannot do business without coltan. It may be possible to stay because the new government has agreed to meet and negotiate. It is also clear that companies from other countries are ready to buy the DRC's coltan if SD does not.

Other aspects of chaos and conflict generate their own risks. Unexploded munitions abound. Staff morale, though recovering, seems brittle.

Pentalemma advisors are therefore presented with a complex and percussive setting in which to advise on a strategy for SD. Thinking beyond concerns about immediate security may be an unrealistic luxury. Even so, several themes must be presented for consideration:

1. The realities of a "hot war" in an area of business operations.
2. The importance of political stability, the implications of new regimes and new ideologies.

3. The value of alternative corporate models as vehicles for equity and peace.

The many issues raised in the scenario include nationalization; the influence of the church; China's role in Africa; inflation and unexploded ordnances.

Scenario Design

Situations of open warfare, whilst extremely dangerous and stressful for any businesses caught in the crossfire, are relatively straightforward in terms of risk management. You keep your head down or you get out. Either way, the sole focus tends to be on safety. It is the immediate aftermath of the war, with the possibility of renewed conflict that can generate more complex dilemmas for management. This is the environment that this scenario set out to explore, with the added complication of a new rebel regime seeking legitimacy.

Business must, therefore, consider not only the obvious fallout from the war, e.g., unexploded landmines and high inflation, but also the politics of a new regime with some radical ideas about the South Kivu economy. The options here are unusually tricky. Particularly tempting is the idea that all is chaos, with planning practically impossible, so that SD should simply continue with as little change as possible. This kind of abdication of management is surprisingly common. Equally straightforward is the proposal to pull out, in view of the new government's hostility toward free enterprise. Both strategies would be costly. To give some examples of this new kind of politics, Patricia Hoenig points to some of her reading of Naomi Klein's writings.

The point here is that engagement with the new government holds the key to improving the situation for Speedy Devices. The offer of a "Peace and Reconciliation" conference hosted by the church is a reasonable one, though it is not clear that this will do much more than kick the can further down the road. Far more creative is the idea that the coltan industry might be reorganized. This stops short of nationalization and builds village-level involvement into a new entity that would sell the coltan to companies such as SD.

Openness to alternative business models is, therefore, one of the more important learning points offered by this scenario. An investment here can produce much more stability, and have important gains for SD. At the same time, the circumstances were often chaotic, representative of a post-war scenario, thereby demonstrating how—even in this situation—there is merit in creative risk management. Finally, this chapter emphasized the importance of engagement with government, however ideologically hostile or transient it might seem.

Questions to Consider

1. What criteria should be considered when determining whether or not to do business with a new political regime?
2. Are landmines, broken transport links and similar side-effects of a "hot war" a legitimate interest for business? In what sense?
3. What would be the risks of association with the kind of a community-based coltan cooperative?

Significant risks in Chapter Three include:

Economic: The lack of stability of the Congolese Franc has led to a demand for payments in US dollars and a sharp increase in the prices of basic goods. SD's original plan to maintain a western presence in the DRC has now become cost-prohibitive.

Ethical: Given that the rebels of the PRSK government now control most of the mines in the area, sourcing coltan in a way that is "clean" has become even more challenging than before.

Political: Since Karen Adams submitted her report, the South Kivu situation has been debated at the UN Security Council. UN troops might not be deployed, but there could be an African Union/European intervention in case of an emergency. SD needs to be strategic, keeping tensions down, but ready to evacuate if necessary.

Social: The rebel government is a coalition between Rai Mutomboki (RM) and Uhuru Mai Mai (UMM), two tribes that were at war only two years ago. SD needs to understand the social dynamics between them to cultivate relationships with the new Administration in South Kivu.

Technological: A new Korean portable dredging and refining plant could potentially transform the cost of coltan extraction in Africa, which could result in new groups of winners and losers.

In such a complex situation, you have other questions to consider. Should SD work with a government that may be corrupt? What if this government collapses? What if PRSK leaders or officials are guilty of war crimes?

SD has a very clear CSR policy that would not allow this kind of engagement. After a long meeting with you to discuss these questions, Keith Poulter sends the below update to SD headquarters:

Pentalemma Inc.: BOARD BRIEFING PAPER
Client: Speedy Devices Inc.
Topic: People's Republic of South Kivu
Author: Keith Poulter, Head of Planning, Africa and Asia

Karen Adams confirms that the People's Republic of South Kivu (PRSK) is under the control of two organizations, the Rai Mutomboki (RM) and Uhuru Mai Mai (UMM).

She says that SD needs to buy coltan at reasonable prices, but must also consider its social responsibility.

Several factors must be considered. Five of these are summarized here:

Security

The South Kivu situation is being discussed at the UN Security Council. Peacekeepers will soon be sent to the region. Fighting continues in the western parts of South Kivu, but the safety of SD staff is not a problem unless the fighting spreads.

Inflation

Local merchants want to be paid in US dollars. Because the only link with the outside world is now through Zambia, the real price for most items is rising fast. Clean water is difficult to obtain. It is becoming harder for SD staff to maintain reasonable living and working conditions.

Explosives

Many landmines, originating in China, have been left behind in South Kivu after the war. Some agricultural workers have been killed or injured. Some of the heaviest fighting was around the coltan fields, so this is an emerging risk. Much work will be required over many years to make these areas safe.

The Rebel Government

This administration is a partnership between RM and UMM. Only a couple of years ago, the two groups were enemies. They are from different tribes. In order to work successfully with this government, SD must build relationships and understand the two groups more deeply.

Technology

A Korean firm has invented a system that could make coltan mining much cheaper. European Electronics, SD's competitor, will test the new equipment in the spring.

These are our current suggestions for your consideration.

Keith Poulter
Head of Planning, Africa and Asia

You have to collect opinions from various experts, including SD managers and Poulter, about what SD should do next.

Karen Adams describes two possible courses of action:

Option A—Prepare to Leave Soon

The new PRSK Government is functioning quite well. Also, it is supported by the church, whose leader, Bishop Mboya, is widely respected. He has offered to mediate between the government, environmental groups and the MNCs. Mboya suggests a new type of coltan mining company, owned by the villages. MNCs would support as investors. The board would include environmental NGOs, the government and the church. This suggestion does not seem realistic.

The government is likely to either nationalize the coltan industry or make coltan buyers pay a hefty tax. Either way, it will be difficult to work with this government. SD and others should plan to leave South Kivu. We should search for new coltan supplies, in Australia, for example.

Option B—Take as much Coltan as Possible

The situation in South Kivu now is really dangerous. During the next 12 months, it will be almost impossible to do normal business. There is a high level of corruption in the new government, which weakens its stability. This means that long-term planning will be difficult. It also means that it will be more expensive to do business.

We must study the costs and risks very carefully. At the same time, we should speed up and get as much coltan as possible while we still can. It will be almost impossible to be sure that all the coltan is "clean." The best we can do is use common sense and keep a detailed record of everything. Security must be our top priority.

Management Toolbox

Can you choose between these two options? Both options have advantages, but they both also include new problems. Serious thinking will be required in order to find creative solutions. After speaking with

Poulter, you agree to ask the master facilitator, Fernando Rodriguez, to lead a problem-solving training session for you and the SD management team. You are confident that Rodriquez can help SD manage the risks successfully. Poulter emails Rodriguez to confirm his availability, and he replies several hours later as follows:

To: Keith Poulter, Pentalemma Inc.
Cc: Pentalemma Response Team
From: Fernando Rodriguez Consulting
Subject: Training next week

Dear Keith,
Thanks for your email.
I understand the urgent situation there. I have booked my flight from London to Kinshasa on Tuesday, transferring and then arriving at N'Djili Airport 17:40 pm local time. As always, let's meet at the arrivals gate.

If you want to give your client good advice, I really think you and your team need to define the problem clearly. We will use the Six Sigma structure of DMAICL (define, measure, analyze, improve, control and leverage). We always find a way forward with these tools.

We will confirm that everybody has the same understanding of the problem. Then we will do some brainstorming. In the end, we will use some decision-making tools to make the final choices.
See you Tuesday then!
Fernando Rodriguez
Master Facilitator

Defining Problems

The Key to Problem-solving Success

Rodriguez arrived yesterday and just before 9:00 a.m. this morning, he begins the first training session with Pentalemma advisors and SD managers.

"If you are lucky," he says, "as soon as you see a problem, the solution will be easy to see. Usually, however, we need an organized process to find a solution. To reach a solution quickly and efficiently, we need to begin with a clear definition of the problem."

Rodriguez tells everyone to think about how they can increase SD's annual profit. Then he tells each person to define that problem, using only one sentence, without showing it to anyone. After that, he asks everyone to read each other's problem definitions. They are surprised to discover that they have very different definitions of the problem. Finally, Rodriguez explains that it is very normal for each person to have a different understanding of the problem until they communicate with each other. An effective problem definition, he says, is specific, objective and concise.

Specific

Your problem definition has to be specific. This means that it must have the key information required in order to formulate a solution.

Objective

It must be objective. If one person says, for example, 'The problem is that we have too many employees,' and then everyone quickly agrees without proper discussion, the result may be counterproductive. To reduce the risk of overlooking important perspectives, the problem definition must have input from a range of people with different views.

Concise

Your problem definition must be short and simple. If the information is unnecessary or has no connection to the problem, it may confuse everyone.

"A good problem definition helps us collect and analyze the necessary information," Rodriguez says.
Examples of good problem definitions:

- Machine A had a 30% rejection rate during March.
- The costs of project B were 10% over budget during Q1.
- Team C is 25 days late reaching an agreement on last year's strategic plan.
- Rodriguez gives workshop teams a series of exercises.

He shows you an example of a poor problem definition. Then team members help each other to change this into an effective problem definition. You check whether or not the definition is specific, objective and concise. Then you check whether the definition answers "what, when, where, how much and why."

> "Always ask questions," Rodriguez emphasizes. "The quality of your result depends on the quality of your questions. Ask who is affected by the problem. Ask whether the problem happened before." This is the team approach to problem-solving. After you have an effective problem definition, your team can think about finding a solution. You can decide what extra information you need. By using this step-by-step approach, you can discover the root cause of the problem. When you have corrected the root cause, the same problem will not occur again. Rodriguez reminds everyone, "Remember, an effective problem definition helps you find at least 50% of the solution."

With help from Pentalemma consultants, the SD teams practice the art of clearly defining problems. Everyone is confident about it. They have learned a lot from Rodriguez. SD can now apply this technique to many of the issues that it has in South Kivu.

Early Media Response

The morning news has a story reflecting Option A, based on remarks from "industry insiders" claiming that SD has decided to pull out of the PRSK.

> Electronics Shares Collapse due to Coltan.
>
> Shares in the world's largest electronics companies fell yesterday because of reports that they will soon leave the People's Republic of South Kivu. They have become convinced that the government will decide to nationalize the coltan industry. Investors fear that China may take over the region's coltan, a mineral essential in the manufacture of electronic goods.

Another news organization has a completely different story, based on other sources who have bought into Option B, the grab-it-while-you-can approach: "Greed rules in South Kivu."

> The coltan mines are busy in the new People's Republic of South Kivu. Ten-year-old boys work from morning to night, mining coltan, a mineral used in electronic devices. Near the mines, gorillas are hunted for food. Big coltan users like Speedy Devices want the mineral at all costs, ignoring the children and ignoring the environment.

When you share these with SD's headquarters, the CEO responds angrily. He calls both Karen Adams and Keith Poulter direct.

"Both of these stories make us look bad. We need to make a clear decision and let everybody know where we stand." He immediately commissions Pentalemma to develop a wider and more creative range of options.

Brainstorming

Using the tools and techniques learned in the training session with Rodriguez, you and your team go back to SD. A long day of intense discussion produces three further options.

Option C—Support Bishop's Conference

The church could mediate. The government, the MNCs and the NGOs have many questions to discuss. For example, is it better for the government to own the coltan industry? How can the landmines be cleared? There are important issues related to education, poverty reduction programs and protecting gorillas.

Most people in South Kivu still follow the church, and almost everyone respects Bishop Mboya. He has offered to support a "Peace and Development Conference." The chance of success may be low, but SD should contribute to this conference. At the very least, SD could continue to do business in South Kivu.

Option D—Invest in Stability

SD does not know whether this government will stabilize, or become weaker and more corrupt. More war in the future is also possible. In any case, SD should tell the government that a tax would be better than nationalization. If the government nationalizes the coltan industry, SD could lose access to the coltan supply (especially if the government makes a deal with China, for example). But, if there is a tax, then some of the funds could go directly to GGC. They could use the money to develop gorilla protection projects and build up the eco-tourism industry. The government and GGC could then work together to make sure that SD's coltan supplies are legal and "clean" (conflict-free).

Option E—Build a Community-based Company

Bishop Mboya's idea to build a new coltan mining company in the area is interesting. His idea is that MNCs could invest in equipment such as pumping plants and safety equipment. The company would be owned by the local villages. The chiefs or their advisors would be members of the board together with people from the government and the church. In this way, the coltan business would be valuable for everyone. It would also reduce government corruption and environmental problems.

The MNCs would not control the company. They would only buy from it. The company could be registered in London or Paris. This would help its image and protect it in case of political problems in the future.

With this plan, the community, the environment and business would all benefit.

Searching for a Win-Win

You now have five options on the table. In the next round of discussions, SD managers state their preferences.

Option A—Prepare to Leave Soon

SD's share price has dropped due to fears that the PRSK Government will nationalize the coltan industry and sell all the coltan to China. People who believe this could happen think it is better to leave now.

They hope SD can buy coltan from the Chinese as well as from other places such as Australia. Australian coltan will be more expensive, but prices will be more stable and risk will be much lower. Those who want to stay think Chinese industry will use all the coltan, so SD should stay in South Kivu as long as possible. Another convincing reason for leaving South Kivu came from the human resources department. They say higher staff motivation would be worth millions of dollars in terms of productivity.

Option B—Take as much Coltan as Possible

The media has many stories about strange characters in the PRSK government. Several SD managers worry that some of these officials will want bribes from coltan buyers. Some also worry that, if the government is corrupt, people will say that SD is paying bribes even if it is not. The lawyers explain that the company can protect itself by keeping detailed, accurate records of everything. The optimists say government officials will be friendly because they are happy for SD to continue creating employment in the country.

Others say that South Kivu is just not safe enough. Security costs are too high and staff motivation is low. At the same time, other costs are escalating because of inflation.

Option C—Support Bishop's Conference

It is important for SD to maintain good relations with both the environmental groups and the new government. If the conference is successful, it will help to protect SD's reputation and share price. It may also enable Bishop Mboya to mediate directly between SD and the PRSK leaders so that normal production can continue. A conference will not solve all problems, however. The risk that experts see is that SD will need to clear landmines from areas where staff must travel. Local employees already have enough to think about—including high prices and the spread of disease—worrying about landmines would be too stressful.

Option D—Invest in Stability

A simple way of remaining in South Kivu is for SD to agree to a new tax on coltan. Investors will not like the extra cost, but SD can expect GGC to be less aggressive. If they are getting part of the funds, to develop eco-tourism, they cannot attack SD's reputation. The PR department wrote that SD should tell everyone about the effects of more expensive coltan. "Let the media have real numbers. If not, they may create a panic about the rising cost of electronics," they say.

Option E—Build a Community-based Company

Several people want SD to fight for the rights of the people of South Kivu and give them new ways to work. They say investors will support a new kind of company that is community-owned and properly managed. They argue that such a company could get a lot of support locally. If so, then the PRSK may not nationalize the coltan industry. Such a course could also reduce corruption as well as make South Kivu much safer for everyone. Those opposed question whether SD really has the resources in this part of the world for such a large project. They believe that SD would probably never recover its investment. They are also concerned that managers and staff are already too busy to take on extra organizing and administrative work. It will be a major task to assess the idea properly.

Now let us fast-forward to consider five hypothetical sets of consequences.

Initial Impact on Reputation

Option A—Prepare to Leave Soon: Electronics Shares Collapse as China Grabs Coltan

Shares in several of the world's largest electronics companies continued to fall yesterday, as they announced their withdrawal from South Kivu, formerly a part of DRC. SD, one of the firms involved, lost 18%

of its value in 24 hours. The pull-out came in response to the new government's statement that it would nationalize coltan production. Even more unsettling for world business was the second part of that statement: China had agreed to buy all of the production through its agents in Zambia. Coltan is vital for the assembly of electronics goods, including phones and PCs. The notion that China will control supplies from South Kivu, which holds one of the few large deposits so far discovered, poses a huge strategic risk for western MNCs.

Option B—Take as much Coltan as Possible—Business as Usual in South Kivu

The coltan mines are busy in South Kivu, now known as the People's Republic of South Kivu (PRSK). 10-year-old boys start work at seven in the morning and continue until dusk. They are digging for coltan, which is vital for the production of electronic goods—my mobile phone, and yours. Outside town, an environmental catastrophe continues as the few remaining lowland gorillas are hunted for bushmeat. The big firms here, like SD, are keeping their heads down. They are paying whatever it takes to whomever they must in order to secure that coltan.

Option C—Support Bishop's Conference—Offering Hope to South Kivu

In a bizarre array that could only happen in this part of Africa, a Church of England cleric has allied himself with hardline Marxists in an effort to plan a stable future for this region. They sat on the stage together, the khaki-clad rebel generals and the purple-robed man of God. Also joining Bishop Mboya's meeting were representatives of the electronics giants who mine valuable coltan here and environmental groups such as the GGC. "We'll see what they've got to say," said Karen Adams of SD. The environmentalists' view was similar: "They say they want to protect the environment here. Well, so do we," said Mark Goodwin of GGC. So, the jury is out on the future for this highly unstable region.

Option D—Invest in Stability—20% Hike in Coltan Tax as New Government Flexes Its Muscle

Electronics giant SD has agreed to pay a 20% tax on coltan extraction in the new South Kivu "People's Republic." It is expected that other players will follow suit and we will all feel the impact on the high street as PC and phone prices respond. A government spokesman claimed the levy would help to build public services and a new sanctuary for the threatened gorillas in the area. GGC, the campaigning group that spun off Andreas Brandt's landmark documentary, is hopeful about this. "We have agreed on a plan with the government and we have no reason to doubt their word," said the organization's Mark Goodwin. "We think the gorillas now have a future." Others are less sanguine: "The whole region remains an unstable war zone," states the Royal United Services think tank in London.

Option E—Build a Community-based Company—Paris-based "Tribal Co-op"

In an astonishing development at Bishop Mboya's South Kivu conference, plans have been announced to set up a new company to run the valuable coltan industry—to be controlled by local tribal chiefs. For reasons of probity and security, the new company will base itself in Paris. "This is not as daft as it sounds," says Karen Adams of SD. "SD will invest in the new company and sit on its Board. But the proceeds from the industry will be controlled by the community. Big environmental players will be involved too." It certainly represents a huge improvement over the anarchic and bloody scramble for coltan that has been the norm here. "We want to cut the criminals out of the loop," one tribal elder told me. "And we want to use our wealth for our own destiny." Cynicism is easy, but it is difficult to fault the ambition here—and with companies like SD behind the deal, it just might fly.

Later Impact on Reputation—Media Reaction

Option A—Prepare to Leave Soon

There has been near panic in some of the financial pages following the announcement about China's proposal to buy up coltan. There are doomsday predictions about strategic minerals, prices, etc. Mainstream media has followed much of this, with slightly more emphasis on the decision by SD and others to pull out. The new government and GGC have made much of their new environmental strategy but this is being treated with near derision by most media.

Option B—Take as much Coltan as Possible

What little reporting there has been on the mining community in the new "state" is of the "business-as-usual" variety. The decision of most MNCs to stay the course does not seem to have attracted much attention. The focus remains on child workers, atrocities and the colorful characters now in government.

Option C—Support Bishop's Conference

Bishop Mboya has dominated the coverage of the conference and its follow-up. The focus has been almost exclusively on security and the quasi-Marxist agenda of the new government. Neither is depicted in optimistic terms. The environmental issue has taken a back seat in terms of what the world reads about.

Option D—Invest in Stability

"Laptop and phone prices to rocket!" The headline from the Times of India was fairly typical. The other line coming from most of the media depicted an eccentric Marxist regime harking back to the 1970s. There were also comparisons with Venezuela. Interestingly, a brief piece in the Economist took the line that this deal would simply not hold and nor would the government. There has been some coverage in the liberal media about the environmental dimension, but again, this has not been extensive.

Option E—Build a Community-based Company

In what was otherwise a quiet weekend for the international media, the announcement of what is being called the South Kivu Co-op (though technically, it will not be a co-operative) received a lot of coverage and support. "Out of bloodshed, comes hope" ran the New York Times Sunday edition. Bishop Mboya received almost adulatory treatment, and the promise of work to save the gorillas was widely welcomed.

General Impact on Firm Value[1]

Option A—Prepare to Leave Soon—A Controlled Exit

COMPANY REPUTATION: Speedy Devices lived up to its name and made a quick exit from the coltan business in South Kivu. This news was not received well by investors in the company who now fear that nationalization may encompass all of the DRC production plants. The share price is down almost 20% from the same time last year. The effect of Mr. Ndagaboy's plan to take control of the coltan extraction and sale of coltan to China had a negative $5,000,000 net value effect.

SALES: Sales were down after pulling out of the South Kivu region. We were able to purchase some of the coltan ore through Lusaka, but not much. As a result, the DRC facilities produced less in aggregate this year than last year. The threat of nationalization caused our removal of business operations from the region and cost, in net value terms, $3,500,000.

RISK MANAGEMENT: As noted elsewhere, the share price fell, and we are now at risk of further drops in the market capitalization of the company as shareholders show growing concern of instability in the region. It is far from clear that SD can secure sufficient coltan from alternative sources either, as other companies scramble for what is available—with a serious threat to continued production. The net value of the pullout was estimated at $2,000,000.

SAVINGS IN SUPPLY COSTS: Perhaps the only benefit to pulling out of the region was to save on the ever-growing supply costs. Shipping from Australia and Zambia—from where for the current year at least we managed to secure sufficient ore, avoids the huge transport costs associated with the DRC's shaky infrastructure. The net value of pulling out of the region was $7,500,000.

STAFF MOTIVATION: The staff were relieved to be relocated to areas of lesser hostility and fighting. We've seen an uptick in recruitment and markedly less staff sickness in our African operations. The net value of moving the SD employees to safety was $4,000,000.

Option B—Take as much Coltan as Possible—Pay What It Takes

COMPANY REPUTATION: Speedy Devices came under intense scrutiny from the group Transparency International with regard to bribes. Although efforts have been made to maintain an accurate account of all financial transactions, SD was cited for involvement in low levels of corruption. This was not good news for the company and its reputation suffered because of it. The net value effect of this bad publicity was negative $1,000,000.

SALES: We have not experienced any instability in our coltan extraction operation due to the new government and the annexing of this part of the country. Perhaps the regime is pleased that locals in the region enjoy stable employment. With no forcible measures taken against us, our sales continued on a level path. We estimate the net value of sales to be $4,000,000.

RISK MANAGEMENT: Despite assurances from internal advisors, the government and international agencies state that the area is not as safe as it once was. Despite the best efforts, it was hard to keep employees motivated. This led to more employee extractions and higher costs for security measures put in place around our production facilities. The net value costs associated with these preventative tactics amount to $2,000,000.

SAVINGS IN SUPPLY COSTS: With massive inflation of the Congolese Franc, local commodities such as fuel have skyrocketed in price, although American dollars will get you a better deal on most products. This has caused our savings estimates to be out of line with actual results. The net value effect from increased costs is estimated at only $2,000,000.

STAFF MOTIVATION: Strife among the warring factions and the murders that took place just miles from some of our production sites led to unrest among our employees. Furthermore, with increased inflation and SD's unwillingness to pay wages in the American dollar, lower motivation among the local workers is to be expected. All told, for the year, there was a loss of $1,000,000 in net value due to staff motivation.

Option C—Support Bishop's Conference—Reconciliation and Development

COMPANY REPUTATION: Locally, the company was seen as acting in good faith with both the environmental groups and the newly established government of South Kivu. Internationally, without nationalization occurring, the Speedy Devices share price remained within normal ranges for the year. The net value of the conference doing its job and saving the reputation and interests of SD was estimated at $35,000,000.

SALES: Thankfully, the company's operations were not nationalized. This was largely due to Bishop Mboya who brokered the existing deal between Speedy Devices and the PRSK leaders. Production continued at a regular pace, since normal operations were allowed to go on uninterrupted. Although the sales in the future may not be so steady, this year added a net value of $5,000,000.

RISK MANAGEMENT: The risk of unexploded ordinance in the region is very high, resulting from decades of warring factions in the DRC and, particularly, South Kivu. Government agencies and NGOs from around the world have started to come to the region to clear the area, but the degree of success cannot be measured, nor is there an established timeframe. Consequently, SD has had to employ its own specialists to clear those areas adjacent to facilities and staff accommodations. Hostilities in the region did not feature despite the constant threat as the conference did its part to quell hostilities. The net value added to SD is zero.

SAVINGS IN SUPPLY COSTS: Savings in supply costs were hard to come by. Fortunately, local dealers were more willing to accept dollars than the deflated Congolese Franc, and this allowed us to maintain somewhat normal operating costs in our supply chain. Elsewhere in the country, our forecast savings were met due to our continued ability to extract coltan and secure our supply routes. Overall savings were $2,500,000.

STAFF MOTIVATION: The staff were worried all year about three major issues, and this affected motivation, sickness and turnover. First, there was a good chance that more fighting would break out. Second, they knew that they might suffer death or injury by one of the thousands of Chinese landmines that were left behind. Finally, they were also concerned about the risk of nationalization should the talks not be successful. All of these stressors put the net value of employee motivation at negative $1,500,000.

Option D—Invest in Stability—Paying the Price

COMPANY REPUTATION: By agreeing to the tax levied on the coltan exports, SD was able to remain in business in South Kivu. This news was not met well by investors; however, as the share price is down and has stayed down 5% since the news was announced. On the plus side, the GGC is now off Speedy Devices's back and we can enjoy extraction without the risk that our reputation is tarnished. The overall net value of this is $4,500,000.

SALES: Sales remained steady as SD continued coltan extraction at current levels. Without a threat from the GGC of retaliatory measures, such as a boycott or negative publicity, demand from our current customers remained even. The benefit of not nationalizing the coltan plants in South Kivu is evident, and sales had a positive net value of $6,750,000.

RISK MANAGEMENT: Certain risks inherent to the region remained throughout the year, namely kleptocracy, corruption and war. The risk that GGC poses was mitigated by their belief that some of the new tax money could be funneled to their operations of eco-tourism and conservation of the wildlife in the area. In essence, a $1,750,000 net value has been added due to the alleviation of these risks.

SAVINGS IN SUPPLY COSTS: Savings to supply costs were as projected. Warring factions and the unstable government did not prevent any of our coltan from getting to the designated region once it was extracted at a now higher price. This equates to a net value for Speedy Devices of $1,250,000.

STAFF MOTIVATION: Local and expat workers were unaffected by the new tax imposed on extractions. Indeed, many of them are happy to still have jobs now that the threat of nationalization has subsided. They do remain wary of the warring factions as well as the unexploded mines and other weapons that pepper the area. Staff motivation can account for a net value of $1,000,000.

Option E—Build a Community-based Company—South Kivu Co-op

COMPANY REPUTATION: The local and national community saw Speedy Devices as an advocate for the locals' rights. SD was rewarded for participating in a community-based extraction company and putting the profits into the hands of the co-chaired new operation. News of this traveled to investors who soon bought up shares of SD on news of the purported synergies this endeavor will bring. The estimated net value is $5,000,000.

SALES: The success of the jointly owned company bolstered the sales of SD. With more coltan being extracted, there was more to purchase and more coltan to sell to the market. Furthermore, it prevented a nationalization program and the development of cartels, which is good for all parties involved. The net value of these increased sales is $3,500,000.

RISK MANAGEMENT: The risk environment arising from this new settlement has changed substantially. Local people feel they have a future to work for, and it seems that confidence in the security situation is building. Already, SD has been able to scale back some of its costly security measures. This equates to a net value of $2,000,000.

SAVINGS IN SUPPLY COSTS: Greatly enhanced security has released big savings in this area. By utilizing the same cargo fleet, supply routes and other such items in the supply chain, Speedy Devices was able to achieve synergies in its logistics operations. The value of these synergies is estimated to be worth $8,000,000 in net value terms.

STAFF MOTIVATION: Locals enjoyed the fact that a portion of their hard work and effort would remain in the community. Everyone involved credits the new company as being a major success story and a model for future operations in the region. The net value added is $1,250,000.

Reflecting on Chapter Three

The Scenario in Context

This final scene in the South Kivu series put the Pentalemma and SD teams right in the middle of a post-conflict zone, i.e., in an area that had very recently been fought over. The premise was that the victor was an alliance of guerrilla groups, who had between them managed to eject central government forces and had now established their own breakaway state.

Somehow, the Speedy Devices operation had survived the fighting relatively intact, and questions now arose about future strategy. The situation remained potentially dangerous, and it was far from certain that the war would not surge back into the area. The rebel government was not firmly established, and yet it was already taking an active interest in the coltan industry, with a nationalization plan. This situation was further complicated by the challenges of a post-war situation and the still influential Gorilla Gorilla Campaign.

Dilemmas and Challenges

The "stay or go" decision was probably the most immediate dilemma in this situation. Naturally, the costs of a pull-out would have been high, and the advice in Keith Poulter's Pentalemma Briefing was that in the event of a new flare-up, African Union or NATO troops had plans in hand to evacuate non-Congolese workers. Having weathered the storm thus far and with this admittedly limited assurance in place, it would have been a rather cautious team that pulled out now.

If SD were to stay, then it had to consider the new government and its rather strident policy toward coltan mining. For all the rhetoric, Karen Adams's reports revealed that the minister concerned was "personable" and that a local church leader had offered to mediate. The fact that GGC remained heavily involved in the region—and had struck some kind of deal with the government—gave weight to the view that these were people with whom one could at least negotiate.

However, SD's concerns extended beyond the awkward politics of the new government. Inflation had pushed costs up, and a new and powerful player had arrived on the scene, in the form of China. Representatives from Chinese companies were proposing to buy all coltan production directly from the government. The local economy was clearly on its knees, with the added challenge of landmines and other war debris littering the area.

Some form of continued presence, therefore, implied an engagement with these issues and parties. A greater understanding of the dynamics of conflict resolution may have helped in putting together a viable approach. The only other alternative would have been to carry on regardless of the risks, in the hope that all will turn out well in the end. There are times when this will work, but it effectively leaves everything to chance—the opposite of risk management.

Note

1. The scenarios in this book are hypothetical, so results are based on plausibility. In general, there are three categories of priorities that I focused on by using John Elkington's concept of the triple bottom line; the idea of balancing financial, social and environmental priorities (Elkington, 1999). As one soon discovers, however, any attempt to balance these inevitably leads to dilemmas. Much of the creativity needed to handle these dilemmas was inspired by Johan Galtung's model of conflict transformation (Galtung, 1996).I studied a wide range of companies that have faced dilemmas in pursuing these priorities, drawing especially on Jane Nelson's research of businesses in high-risk environments (Nelson, 2000). Adam Kahane's Mont Fleur scenarios informed the storyline by providing a firsthand review of the planning process inside companies operating in conflict areas (Kahane, 2012). On the whole, extrapolating from these cases, the results for the majority of the stakeholders are usually unsustainable when their decisions only serve a minority of their stakeholders. The results are also negative when decisions are delayed or not implemented. The results tend to offer more lasting value when decisions satisfy the needs of most of the main stakeholders reasonably well over longer timeframes.Finally, to apply the lessons learned from Elkington, Galtung and Nelson to practical management challenges, Ed Freeman's work on Managing for Stakeholders provided helpful examples of how everyday management dilemmas can be reframed as resource issues between stakeholders (Freeman, 2007). The scenarios have drawn on lessons learned from these and other scholars and practitioners to create the heuristic process used for developing plausible outcomes to the various dilemmas presented in this book.

Further Reading

Elkington, J. (2002). *Cannibals with forks: the triple bottom line o- 21st century business.* Oxford: Capstone.

Freeman, R. E., Harrison, J. S., & Wicks, A. C. (2007). *Managing for stakeholders: survival, reputation, and success.* New Haven, CT: Yale Univ. Press.

Galtung, J. (2012). *Peace by peaceful means peace and conflict, development and civilization.* Oslo: Prio.

Kahane, A. (2012). *Transformative scenario planning working together to change the future.* San Francisco, CA: Berrett-Koehler.

Nelson, J. (2000). *The business of peace: the private sector as a partner in conflict prevention and resolution*. London: International Alert.

Conclusion

Highlights of Chapter One

Some of the more important themes that the Chapter One scenario was designed to highlight are considered below, in the context of the options which were available. It is hoped that this will serve both as future reference and also as a further explanation of the linkage between the dilemmas and outcomes enshrined in Pentalemma.

A critical concept in this scenario, particularly given the instability of the region, is that of cooperation. This extends not only to the local community and governing authorities, but also to competing businesses that nonetheless have shared risks and interests. You will have found that the three options (C, D and E) produce the best results for SD. Of these, it is Option E which explicitly uses the notion of comprehensive local engagement with all parties as a starting point; and indeed, this selection, which envisages establishing a secure basis for future business, generates rewards in areas such as risk management and company reputation. Option C takes up the worthy ambition of information gathering and assessment, but perhaps implies delays which are difficult to justify given the urgency of the situation, and Option D secures more modest progress in important areas.

Regarding morality and profit, few would dispute that the exploitation of children in the manner that has been commonplace in South Kivu should be eradicated. Furthermore, the Dodd-Frank Act makes it clear that any company with shares traded on the New York Stock Exchange needs to demonstrate that it has taken measures to avoid association with this and other abuses. The certain and immediate means of ensuring a clean reputation in this area would have been for SD to select Option A as its strategy—pulling out of the area entirely. Pursuing this option would have revealed that sales dropped due to dislocation and that shareholder confidence collapsed. Relocation costs would have been high, with little prospect of government subsidy. Alternatively, Option B—business as usual, would effectively have sidelined the problem while pretending that Kids In Crisis had the issue under control.

The shrewder approach would have been to recognize that this was an issue with less resonance in the DRC than (say) in the USA, which it might be possible to address once more pressing problems had been mitigated. This more mature and sophisticated attitude is open to criticism but would be likely to bring along all of the players in the process, even though that might take time. Any of the other three options would facilitate this, and all of them would allow for taking reasonable "Dodd-Frank" measures in the meantime.

The optimal strategy for your team to have adopted in this scenario was Option E. It is this approach that offers an imaginative future for the region, taking as its first precept the idea that the local priorities of security and poverty should also be those of the businesses working in South Kivu. Precursor conditions for peace are what count under such circumstances, with seemingly abstract concepts such as good local governance recognized as fundamental building blocks. The more immediate suggestions set out in Option D are all very sensible and will generate modest payback in areas such as staff motivation. But the big opportunity lies in taking a global view of the risks faced by the community as well as by the company and beginning to work collectively to deal with those. The child labor question and others, such as environmental protection would all take their place within that framework and be easier to deal with as a consequence. A comprehensive risk overview is imperative for a company operating in a conflict situation. Focusing on one highly emotive aspect of the situation is less likely to bring lasting benefits.

Highlights of Chapter Two

Some of the important themes that this scenario was designed to highlight are considered below, in the context of the range of options available at the time. It is hoped that this will serve both for future reference and also as a further explanation of the linkage between the dilemmas and outcomes enshrined in Pentalemma.

Arguably the most useful element in the plan outlined within Option E is in some ways a by-product, in that the development of an eco-tourist industry to protect the gorillas actually creates a broader local economy. The option envisages SD co-branding with this initiative and putting up hard cash through its support of an anti-poaching unit. Option D is similar in that SD would offer considerable support to the environmental lobby, but it lacks the integration with the emerging new economy that Option E clearly identifies. Why should this matter to companies whose real interest is buying coltan at a good price? A more balanced and larger local economy will offer increased prosperity and employment for local people. This acts as an important normalizing influence in a community, with less scope for criminal or rebel groups. The payoffs for SD will include much better staff morale, lower risk management costs and, of course, an enhanced reputation. It is this imaginative and multi-dimensional strategy that provides the optimum selection for this scenario, and the key to it is in long term economic development.

One of the interesting twists in this situation relates to the attitude of SD's own staff on site. Karen Adams immediately signals her sympathy for the environmental cause—which it would be possible to regard as slightly disloyal. In fact, the attitude she expresses simply mirrors that of many people in other parts of the world, as evidenced by the popularity of the film and the support for the GGC. In this sense, staff opinion can sometimes serve as a bellwether for companies prepared to listen. This is especially true when those on the frontline are more acquainted with local conditions than senior management. It can be a double-edged sword: Karen Adams might have had a personal preference for Option D, which whilst generous in outlook, would have achieved little of long term benefit to SD. The uptick in reputation and staff motivation would have been more than offset by poorer metrics in other areas. The views of staff are certainly to be given due place, but it is for management to set these in the context of the broader and longer-term picture available to them.

The sudden escalation of a seemingly low environmental risk lies at the heart of this situation. Public outcry, especially when fueled by environmental concerns, can be one of the most potent threats to a company's reputation. Indeed, reputation is one of the more volatile indices reflected in the managers' reports in this scenario. The temptation might have been to ride out the storm, choosing Option B. Yet the sales slump and high-risk costs associated with this strategy make it a poor choice. This type of risk needs to be identified and managed. It calls for active horizon scanning and the acknowledgment that reputation is, in the modern world, a crucial commodity.

Highlights of Chapter Three

Some of the more important themes that this scenario was designed to highlight are considered below. Many factors in this scenario were beyond the control of the company. Both the DRC government and the rebel government are major stakeholders. It is not known whether the rebel state of PRSK will survive, or whether violence will make it impossible to do business in the DRC.

A decision to pull out would have been understandable in this context, but would also have introduced different risks—such as a share price slump. Assuming that SD did not pull out, the real lesson here was one of engagement. Companies that decided to continue with their business oblivious to the other issues swirling around them would have seen supply and risk costs escalate, as the environment failed to improve. In a post-war situation, the population will be desperate to return to normality. The government will be trying to rebuild the economy. Problems can be solved, and advice on dispute resolution is available. It is far better for business to engage in that process and help it along, with a seat at the table and a voice that might be heard.

It follows from the above that the three options which envisaged an active and committed SD presence in South Kivu for the foreseeable future represented the better choices for readers. One of the subsidiary dilemmas of such a course is that of working with a government with questionable legitimacy and a seemingly hostile agenda. Businesses, if they are to operate in any given society, must deal with the existing authorities. If a government is established to be, for example, culpable in war crimes, then that makes a good case for withdrawal. But continued presence must mean engagement. In many cases, a government's announced position will be subject to change following negotiations with interested parties. Global companies will invariably have something positive to offer a new administration, and that is very much the case in this scenario.

There was a compromise option available to the teams in this scenario which essentially struck a deal with the government entailing a tax levy on the coltan industry, much of it ring-fenced to environmental protection measures. This was a perfectly reasonable approach to take and would have led to a reasonable outcome for SD. The scenario posits reduced risk-management costs but continuing supply headaches over the ensuing 12 months.

The bolder move would have been to sign up to the alternative business model enshrined in one of the other choices. This proposed the establishment of a new company that would control extraction and sell the product to SD and others. Business, government and environmental interests would be represented on the Board, which would be controlled and owned by local people. In this way, SD would be able to contribute its expertise to a new concept. A longer-term plan such as this, working to keep the production of coltan away from criminal elements, would be to the advantage of manufacturers such as SD. The security of supply would be greatly enhanced, as would company reputation. This is clearly the best of the available choices.

Other Books by This Author

Multi-Hazard Disaster in Japan

Learn about disaster management following an earthquake in Kyushu. By advising Brighter Brand Japan, a company with activities around the country, readers acquire logical thinking skills, and the ability to distinguish between long-term strategies and short-term tactics.

Coffee and Conflict in Colombia

You and your team advise Caffeine Supply, a global coffee exporter. The company has a successful history, but the overall industry now faces decline at home and increasing competition from abroad. You will learn about fair trade and the basics of negotiation in a cross-cultural context.

Mining Conflict in the Philippines

Welcome to Mindanao, an island in the south of the Philippines. You will help Massive Metals, a foreign mining company, to manage its stakeholders. You will learn to use problem-solving tools to resolve environmental and political issues.

Notes to Self

Chapter One

1. Who are the main stakeholders?

2. What are the stakeholders' main interests?

3. Which issues do the stakeholders disagree about?

4. Which of the above issues is the most critical one?

5. If the most critical issue is not resolved, what might happen to the stakeholders and your client?

6. Which option do you think is the best one for your client to choose?

7. How will the different stakeholders benefit from choosing the option you suggest?

Chapter Two

1. Who are the main stakeholders?

2. What are the stakeholders' main interests?

3. Which issues do the stakeholders disagree about?

4. Which of the above issues is the most critical one?

5. If the most critical issue is not resolved, what might happen to the stakeholders and your client?

6. Which option do you think is the best one for your client to choose?

7. How will the different stakeholders benefit from choosing the option you suggest?

Chapter Three

1. Who are the main stakeholders?

2. What are the stakeholders' main interests?

3. Which issues do the stakeholders disagree about?

4. Which of the above issues is the most critical one?

5. If the most critical issue is not resolved, what might happen to the stakeholders and your client?

6. Which option do you think is the best one for your client to choose?

7. How will the different stakeholders benefit from choosing the option you suggest?

PETER LANG
PROMPT

Peter Lang Prompt is a new product that offers our authors the possibility of publishing their research in its intended form, outside the traditional length of a standard book or a journal article. It is produced within a faster production time, without missing the high standard of scholarly publishing. This concise model gives researchers the chance to publish their time-sensitive research, open a forum for debate, and create a greater impact.

For further information, please contact:

Peter Lang Publishing, Inc.
Acquisitions Department
29 Broadway, 18th floor
New York, New York 10006

To order, please contact our Customer Service Department:

peterlang@presswarehouse.com (within the U.S.)
orders@peterlang.com (outside the U.S.)

Visit our website: www.peterlang.com

Prompts include:

Claudia Aburto Guzmán, *Poesía reciente de voces en diálogo con la ascendencia hispano-hablante en los Estados Unidos: Antología breve* (978-1-4331-5207-8)

Janet Farrell Leontiou, *The Doctor Still Knows Best: How Medical Culture Is Still Marked by Paternalism* (978-1-4331-7322-6)

Josiane Ranguin, *Mediating the Windrush Children: Caryl Phillips and Horace Ové* (978-1-4331-7424-7)

Dylan Scudder, *Coffee and Conflict in Colombia: Part of the Pentalemma Series on Managing Global Dilemmas* (978-1-4331-7568-8)

Dylan Scudder, *Conflict Minerals in the Democratic Republic of Congo: Part of the Pentalemma Series on Managing Global Dilemmas* (978-1-4331-7561-9)

Dylan Scudder, *Mining Conflict in the Philippines: Part of the Pentalemma Series on Managing Global Dilemmas* (978-1-4331-7632-6)

Dylan Scudder, *Multi-Hazard Disaster in Japan: Part of the Pentalemma Series on Managing Global Dilemmas* (978-1-4331-7530-5)

Angela Williams, *Hip Hop Harem: Women, Rap and Representation in the Middle East* (978-1-4331-7295-3)

Ivan Zhavoronkov, trans., *The Socio-Cultural and Philosophical Origins of Science* by Anatoly Nazirov (978-1-4331-7228-1)

www.ingramcontent.com/pod-product-compliance
Ingram Content Group UK Ltd.
Pitfield, Milton Keynes, MK11 3LW, UK
UKHW021828100926
8556IPUK00003B/214

9 781433 175619